Guide to Modern Communication Technologies

Kusal Epa

MSc(Loughborough), CEng(UK)
MIET, MIEEE

Preface

Rapid advances in communication technology in the recent past have introduced a whole array of new communication technologies. Digital communication technologies are now offering value added features at higher speeds with better qualities and are gradually replacing analogue communication technologies. Number of new and emerging communication technologies that are emerging have sometimes confused the general public. The aim of this book is to give the reader a basic overview of many emerging and currently used communication technologies. Next few years will see the widespread use of these technologies and it will be useful for the public to have knowledge of these new technologies. Knowledge of these technologies would benefit the reader to make better choices and adapt to the emerging world of modern communication.

Kusal Epa
MSc (Loughborough), CEng (UK)
MIET, MIEEE

January 2013

Contents

Introduction

Advances in communication technologies in the recent past have revolutionised the way people communicate with each other. Communication technologies in the past were primarily of analogue type. Digital communication technologies have improved the way people communicate and information is distributed. Digital communication technologies can provide value added services, higher amount of content at a faster rate with better clarity and quality. People living in most parts of the world now have the ability to get in touch with each other in an instant and news about events happening around the globe are spread around rapidly due to the use of modern communication technologies.

Few decades ago the primary form of communication was by letter and it took days for information to arrive. Telephones were available only for a few; expansion of the telephone network changed this scenario enabling more people to communicate with others instantly. Facsimile machines enable images and documents to be sent across phone lines.

Introduction of mobile phones made the paradigm shift in the way people keep in contact enabling them to be reachable almost anywhere anytime. Places where fixed line telephones were not viable are now reachable on mobile phone networks. Satellite phones now enable communications from virtually any place on earth.

Analogue mobile phone networks have now been replaced by second generation digital networks. Third generation mobile phone networks have already been deployed in most parts of the world. Next generation mobile networks like LTE (Long Term Evolution) have already been deployed in some countries for higher value added mobile phone services. Femtocells, which are a form of a miniature mobile radio base station, arc starting to be deployed in homes and workplaces for better mobile phone and mobile broadband access within those premises.

Internet has enabled people to access a load of information from all over the world. Dial up modems were initially used to connect to internet. ADSL broadband services have enabled access to internet throughout the day at faster speeds. Wireless mobile broadband services like HSPA makes it possible to connect to internet on mobile phones, on a computer or tablet from wherever mobile network coverage is available. WiMax is another high speed wireless access technology.

Apart from access to information, internet is now being increasing used for voice communication using VoIP (Voice over Internet Protocol). VoIP enables people with internet access to talk virtually free of charge even across countries. SIP (Session Initiation Protocol) is another method which uses internet for phone calls and does not even need a computer when a SIP device is available for communication.

Radio was an important source of information and entertainment for many decades. FM radio has enabled hi-fi quality music to be broadcast. DAB (Digital Audio Broadcasting) services are starting to give additional features with higher quality radio broadcasting. Satellite radio services enable people to listen to a wider selection of radio channels from around the world.

Analogue Television services are now migrating to digital television services which can offer more channels with better quality pictures, some countries have already announced the dates when analogue television channels will be completely switched off. Satellite television services and cable TV offer hundreds of channels from around the world. IPTV services are delivered using the telephone and cable networks.

Wireless technologies have enabled devices to be connected using radio waves without attaching cables. Bluetooth enables mobile phones to be connected to a mobile phone headset for a hands free phone conversation. Wi-Fi enables computers and peripherals can be connected by wireless means; it is no longer needed to attach the computer to a broadband modem to access internet using cables when Wi-Fi is available. Cordless phones enable one to talk using the home phone line with the flexibility to walk around the home or garden while engaged in a call.

For high speed data connectivity ISDN was initially used which facilitated digital data connections over the phone network. ADSL and ADSL 2+ services now offer affordable internet connectivity for homes and workplaces. Leased lines can offer dedicated data links mostly for commercial establishments. Metro Ethernet is another method that provides high speed connectivity to building in metro areas.

Fibre optic cables can facilitate the transfer of huge amount of data using light waves for data transmission. High capacity fibre cables and submarine fibre cables enable millions of telephone calls and data connections across the world. FTTH (Fibre to the Home) connections where each home will have a fibre optic connection will enable very high speed internet access and many other services.

GPS (Global Positioning System) which is a satellite based navigation system have enabled people to have location information enabling them to easily know their exact locations and plan their journeys with ease. GPS devices with digital maps in vehicles are now used in many countries for navigation and have replaced printed road atlases.

Trunked Radio services provide communication services for emergency and utility services using shared radio frequency bands.

The following chapters of this book give an overview of these new and emerging communication technologies.

1

Cellular Mobile Phone Service

Mobile phones also known as cellular phones enable users to make and receive telephone calls anywhere within a mobile network coverage area. The popularity of mobile phones was due to the ability to use the phone virtually from anywhere while being mobile and not only from a fixed location.

Mobile phone networks are operated by different mobile phone operating companies and the coverage can vary for different networks. A mobile phone is connected by radio waves to the nearest radio base station of the mobile network. If radio base station coverage of a particular network is not available in a given area the mobile phone cannot be operated. Mobile phones registered with a particular mobile network can only connect to radio base stations of that mobile phone network.

Cellular concept

Mobile phone networks are also known as cellular mobile phone networks since the networks comprise of a number of cells or areas each served by a separate mobile radio base station. Mobile phones within each cell area are connected to the radio base station of that cell area.

1

Mobile phone radio base stations are known as Base Transceiver Stations (BTS), Radio Base Stations (RBS), or Node Bs in 3G. Base stations are connected to a Base Station Controller (BSC)/Radio Network Controller (RNC) using microwave, cable or fibre links. BSC/RNCs are in turn are connected to a Mobile switching centre (MSC). A call taken from a mobile phone is connected to another phone line at the mobile switching centre. If the call is to a different network the call is routed through to the Public Switched Telephone Network (PSTN).

Mobile phones will always emit the minimum transmitter power required to reach a radio base station. As a mobile phone user gets away from a radio base station the mobile phone will automatically increase its transmitting power in order for its signals to reach the radio base station.

Frequency reuse

The concept of frequency reuse is a feature in cellular mobile networks. Hundreds of cellular mobile phones in use make it necessary to re-use the limited number of radio channels many times in different areas without causing interference to each other. Numbers of radio frequency channels used to connect mobile phones to the radio base station are reused in different cells of different areas in a mobile phone network. In a mobile phone network, frequency planning is done in such a way that adjacent cells do not use the same frequency channels to prevent interference. In 3G UMTS and CDMA mobile systems CDMA modulation is used and the same wideband frequency channel can be used in adjacent cells but with different codes embedded in them prevent any interference.

TDMA

TDMA (Time Division Multiple Access) is the multiplexing scheme used in digital cellular mobile phone networks. In TDMA a single radio channel can be used by many users at a given time. Different time slots are allocated to each user to communicate enabling them to share the channel. Speech signals are digitised and digital data are sent as short bursts. In GSM each channel has 8 time slots enabling 8 users to share a radio channel at a time. Each user sends or receives data in the allocated time slot for user. Switching each user to their allocated channel and time slot to communicate are done very fast with each time slot having a time of about one thousandth of a second. Users do not even notice that the radio channel is shared by many others.

CDMA

3G UMTS Mobile phone systems and CDMA mobile phones, in addition to TDMA also use CDMA system which is explained in a separate chapter.

Handover/Handoff

A mobile phone not only enables one to make or receive a call from anywhere within a mobile coverage area, it also enables a call to be maintained while moving from one cell area to another.

A mobile phone constantly monitors signals from radio base stations around the area. While it is connected to the nearest radio base station it also monitors signals from other neighbouring radio base stations in the area. When a mobile phone user is travelling while engaged on a call and is going away from a particular radio base station but closing in to another base station the mobile phone connection is transferred through to the nearer base station in another frequency in an instant without the user even noticing. This happens without even disrupting the current call. This is known as handover or handoff.

In 3G mobile phone networks which use the CDMA scheme a handover procedure known as soft handover is used. In soft handover a mobile phone while engaged in a call and moving away from a radio base station will first make a connection with the nearing radio base station while still maintaining contact with the earlier radio base station. After a while the initial connection to the radio base station is broken and all communication will take place through the newly connected radio base station.

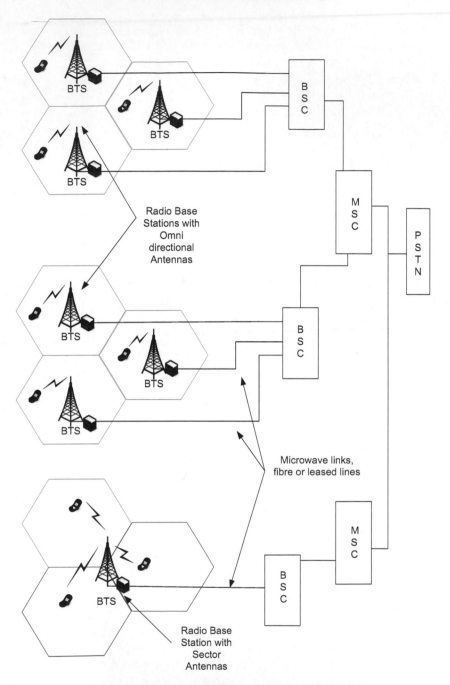

Radio Base
Stations with
Omni
directional
Antennas

Microwave links,
fibre or leased lines

Radio Base
Station with
Sector
Antennas

Basic configuration of a Cellular Mobile Phone Network

BTS – Base Transceiver Station, also known as RBS (Radio Base Station) or NodeB in 3G.

BSC – Base Station Controller also known as RNC (Radio Node Controller)

MSC – Mobile Switching Centre

PSTN – Public Switched Telephone Network

2

Evolution of Cellular Mobile
Phone Technologies

Cellular Mobile phone technologies have evolved and developed a long way from the first analogue mobile phones to the current digital mobile phones. Different technologies are identified as belonging to different generations beginning from first generation to the latest fourth generation.

1G - 1st Generation Networks

Speech signals were transmitted in analogue signal format in these networks. There were several types of analogue networks that were in use.

TACS – Total Access Communication Services. E-TACS or Extended TACS were also in use

AMPS – Advanced Mobile Phone Systems. Mostly used in Australia, USA, South America and few other countries.

The coverage provided from analogue network radio base stations were usually much larger than that of digital radio base stations but the capacity of phone calls that could be provided at a time were low. 800 and 900 MHz range frequencies were used for these networks.

2G - 2<u>nd</u> Generation Networks

These networks were digital networks where speech signals were transmitted in digital signal format. Many value added features like SMS were introduced in 2G networks.

2G technologies can be divided into TDMA-based and CDMA-based standards depending on the type of multiplexing used.

GSM

GSM is TDMA based and was originally developed for Europe but is used worldwide. GSM is the abbreviation for Global System for Mobile communications. GSM is the most widely used network throughout the world with over 1000 networks in operation in different countries.

GSM operates in the 900-MHz and 1800-MHz bands in Europe and Asia and in the 850-MHz and 1900-MHz band in the US Continent.

IS-136 (D-AMPS)

This is the digital version of AMPS networks and is TDMA based. USA and a few countries had this type of network.

IS-95 (CDMAOne)

CDMAOne is a CDMA based, (commonly referred as simply CDMA in the USA) mobile phone network, used in the Americas and parts of Asia

PDC

PDC is a TDMA based 2G network used exclusively in Japan

2.5G

2G systems had limited data communication capability which was overcome in 2.5G systems with packet switching technology.

GSM with GPRS and CDMA 20001X are known as 2.5G technologies.

2.75G

GSM with EDGE enhancement is known as a 2.75G technology.

3G – 3ʳᵈ Generation Networks

3G or 3rd Generation phone networks can be classified in to a number of different categories

3G UMTS (W-CDMA)

3G UMTS (Universal Mobile Telephone System) is based on W-CDMA (Wideband Code Division Multiple Access) radio access technology.

CDMA2000 1X EV-DO and CDMA2000 1X EV-DV

This is a development of 2G CDMA standard IS-95. CDMA2000's main promoters are in the Americas, Japan and Korea.

WiMax

WiMax is categorised as a 3G technology although promoters of WiMax claim that it has higher capabilities than 3G and nearer to 4G standards.

TD-SCDMA

A system developed in China.

3.5G

HSDPA is an evolution and improvement of the 3G UMTS mobile phone system which gives higher data transfer speed and is commonly known as a 3.5G technology.

3.75G

3G Mobile phones networks with HSUPA (High Speed Uplink Packet Access) capability are identified as 3.75G systems.

3.9G

LTE (Long Term Evolution) is considered as a 3.9G technology (also known as 3.99G), but many mobile phone network operators term LTE as a 4G technology for marketing purposes.

4G – 4th Generation Networks

LTE-A (LTE Advanced) system is expected to fulfil the 4G standards. WiMax release two version is also expected to be a 4G technology.

3

GSM

GSM is the abbreviation for Global System for Mobile communication and is the most widely used cellular mobile communication system in the world. It is a 2nd Generation (2G) digital mobile communication system. GSM was originally designed for use in Europe but is now used all over the world. There are over 4.5 Billion GSM users in the world which is more than 80 per cent of the worldwide mobile phone users.

SIM card, SMS and Roaming were first introduced to the mobile communication world in GSM networks.

Most GSM systems operate in the 900 MHz band. 1800 MHz band is also used for urban areas. In USA GSM uses 850MHz and 1900MHz bands. A pair of frequencies 45MHz apart in 900MHz band and 95MHz apart in 1800MHz band is used for uplink (mobile terminal to radio base station) and downlink (radio base station to mobile terminal) connections. The type of radio frequency modulation used in GSM is known as GMSK (Gaussian Minimum Shift Keying) which is a form of Phase Shift Keying modulation. Each GSM radio frequency channel is 200 kHz wide and can carry up to eight simultaneous telephone calls using TDMA (Time Division Multiplexing Access) scheme.

GSM signals from a radio base station can communicate with a mobile terminal up to a distance of 35 kilometres. Mobile phones do not communicate with a radio base station if it is more than 35 kilometres away even if signals are present due to synchronisation issues.

GSM supports data communication speeds of up to 9.6kbps (without the GPRS and EDGE enhancements). GPRS and EDGE are enhancements in GSM system which facilitates faster data communication.

In GSM a VoCoder (Voice Encoder) is used for speech encoding. Speech signals are encoded in to a 13kbps channel using these Vocoders.

3G UMTS is the 3rd Generation (3G) mobile communication system evolving from 2G systems like GSM. Most 3G mobile phones will initially be of dual mode type supporting both GSM and 3G UMTS technologies.

4

3G UMTS

3G UMTS is the abbreviation for 3rd Generation Universal Mobile Telecommunication System. It is commonly known as 3G mobile. 3G mobile networks provide higher call handling capacity and higher data communication speeds than 2nd Generation (2G) mobile communication networks such as GSM or CDMA. A variety of value added features like MMS (Multimedia Messaging Services) and Video calling are also available in 3G mobile networks.

UMTS support a data transfer speed of up to 384kbps (without HSPA enhancement). HSPA (High Speed Packet Access) is an enhancement to the 3G UMTS to facilitate faster data transfer.

3G Mobile Phones

A 3G mobile phone is needed to use 3G mobile phone services. Almost all 3G mobile phones are of dual mode type supporting both 3G and 2G. 3G networks are usually deployed on top of 2G networks and whenever 3G coverage is not available mobile phones will switch to 2G mode and connect through the 2G network.

USIM

3G UMTS phones use a USIM (Universal Subscriber Identity Module) card. Most USIM cards can be used in GSM phones as well where the connection will then be established through a GSM network.

WCDMA

Radio access part of the UMTS is known as WCDMA (Wideband Code Division Multiple Access). CDMA is the radio frequency modulation scheme used here. A radio frequency channel in 3G UMTS is 5 MHz wide.

3G Radio Access Frequencies

Most 3G UMTS networks use the 2100MHz radio frequency band. 850MHz band is also used in USA and Australia. Some mobile phone networks have started to use the 900MHz GSM band for 3G UMTS networks. A pair of frequencies 190MHz apart is used for uplink and downlink transmissions (in the 2100MHz band)

3G UMTS networks facilitate faster data transfer speeds than 2G and provide more capacity for calls. Initially both 3G UMTS and 2G networks will both coexist. It is expected that 3G networks will gradually replace 2G networks in future.

5

LTE

LTE (Long Term Evolution) is the next version of mobile communication system evolving from 3G UMTS (Universal Mobile Telecommunication System). LTE will enable mobile data transfer speeds many times faster than 3G UMTS mobile communication systems. Commercial LTE networks are now being deployed by Mobile phone network operators.

LTE is based on an all-IP platform where only packet switching will take place and there will not be any circuit switching. All data including voice calls will be converted to IP packets for transport over air in LTE. VOIP technology is used to provide voice phone calls in LTE.

LTE will employ multiple-input multiple-output (MIMO) antenna technology. In MIMO, cellular radio base stations will have additional transmit and receive antennas. Mobile phones will also have one transmit antenna and up to two receive antennas.

Peak Data download speeds of 326Mbps and upload speeds of 86Mbps are expected in LTE when using 20MHz wide bands with 4X4 antennas (4X4 – 4 transmit and 4 receive antennas).

Optimal cell size of an LTE Radio Base Station will be around 5km while 30km cells with reasonable performance and 100km cells with acceptable performance would be supported. Unlike 3G UMTS networks, in LTE radio frequency bands of different bandwidths ranging from 1.5MHz to 20MHz can be used in a wide range of radio frequencies that will be available. Both FDD (Frequency Division Duplex) which require a pair of frequencies for uplink and downlink and TDD (Time Division Duplex) which uses a single frequency band for both uplink and downlink in different time slots would be features of LTE radio access network.

LTE networks already deployed use different frequency bands varying from 700, 800, 850, 1700, 1800, 1900, 2100, 2300, 2600 MHz bands depending on the country and mobile network operator.

OFDM (Orthogonal Frequency Division Multiplexing) will be the modulation scheme of the Radio Access Network for the downlink while SC-FDMA (Single Carrier-Frequency Division Multiple Access) will be the modulation scheme for the uplink since it has less battery power consumption in hand held devices. QPSK, 16QAM and 64QAM data modulation technologies are supported in LTE.

LTE is commonly termed as a 3.9G technology but many mobile network operators term it as a 4G (4th Generation) technology for marketing purposes. LTE devices will initially have the capability to fall back and operate in 3G/2G modes when LTE network coverage is not available.

LTE-A (LTE Advanced)

LTE Advanced system which will have further enhancements to the LTE system is believed to fulfil the actual 4G standard in mobile communication systems. Most of the enhancements from LTE to LTE Advanced networks are expected to be only software upgrades. Peak Data transfer speeds of 1Gbps for downlink and 500Mbps for uplink are expected to be features of LTE Advanced

6

Mobile Phone Features

Mobile phones have different value added features which are quite useful for various situations. Some of the commonly used features are explained below.

SIM Card

GSM networks introduced the use of SIM cards in mobile phones. SIM is the abbreviation for Subscriber Identity Module. It is a portable memory chip in the shape of a tiny rectangular card. Most digital mobile phones need a SIM card to be inserted in the phone for it to operate. In most phones it is inserted beneath the battery of the phone. Some new smartphones and tablets require smaller versions of the standard SIM card known as micro SIM or nano SIM. The SIM card holds in it personal identity information, cell phone network and phone number related information, phone book and saved contact names with numbers (if saved to SIM card), text messages and other data   SIM card carries the mobile phone number with it, which is transferred when inserted to another mobile phone. SIM cards used in 3G UMTS networks are commonly known as USIM cards. Dual SIM phones allow two SIM cards to be inserted and the phone has two phone numbers available at all times.

When travelling abroad it is useful to insert a prepaid SIM card of the visiting country to the phone and have a local number of the visiting country for communication. A different SIM card can be inserted and operated only if the phone is not SIM locked or network locked. Alternatively roaming service can be used where the original phone number is used in the visiting country.

A SIM card

SIM backup

Information saved in the SIM card like contact list and short(text) messages can be backed up and saved to a computer using appropriate software which is usually supplied with a phone and connecting the mobile phone to a computer.

Mobile Phone Access Codes

SIM lock

Some mobile telephones, specially when they are issued at subsidised rates, may be SIM locked. SIM locking is done by phone companies to prevent the mobile phone being used with a different SIM card.

Mobile phones can either be SIM locked to a particular mobile phone network or to a specific SIM card.

If a phone is locked to a particular SIM, a different SIM card will not work with that phone. If a phone is locked to a particular mobile phone network, only SIM cards of that mobile phone network will work in the phone.

To unlock SIM locked phones, a sequence of codes have to be entered on phone keypad, unlocking codes have to be obtained usually from the company that issued or locked the phone.

Security/Lock Code

Most phones are equipped with one or more codes to protect it against unauthorized use. The security/lock code is usually a 5 digit number and pre-set to 12345 on some phones. This code can be changed to any user desired number. When the security/lock code function is enabled the phone is set to request the code before use.

The security code, which protects the phone itself, differs from the PIN code, which protects the SIM card.

PIN Number (Personal Identification Number)

Mobile phone SIM cards have the feature where a PIN function can be enabled. When PIN security function is enabled the user is required to key in the PIN to unlock the keypad. If an incorrect PIN is entered three times successively the SIM is blocked. In most SIM cards the PIN is a 4 digit number and default PIN is set to 0000. The PIN also can be changed to a user desired number.

PUK (Personal Unblocking Key)

When the SIM is locked due to three successive incorrect PIN entries a different code to unlock called PUK (also known as PUC – Personal Unlocking Code) is needed. It is a 4-8 digit number and is usually supplied with the SIM card. If for the next 10 attempts the wrong PUK is entered the SIM is permanently blocked and a new SIM card may need to be obtained. PUK may be supplied with the SIM card or can be obtained from the mobile phone network provider or agent who sold the phone.

PIN2/PUK2

For certain network functions a PIN2 code may also be supplied with SIM card. PIN2 provides a second layer of protection against unauthorised use. If PIN2 is locked due to incorrect entry, PUK2 is needed for unlocking.

Barring Passwords

Barring passwords can be set in most mobile phones to restrict access to certain features and functions like making outgoing calls, making international outgoing calls or to bar incoming calls, to bar international incoming calls etc. These features are useful in different situations, to enable these features certain codes need to be entered in the phone keypad.

IMEI

IMEI stands for International Mobile Equipment Identity. It is a unique 15 digit number used to identify a mobile phone or device in a GSM or 3G Network. It is similar to the serial number of a phone. IMEI number is usually indicated in the mobile phone in a white sticker beneath the battery, mobile phone container box also has the IMEI indicated with a barcode in a white sticker. In most phones IMEI number can be obtained by keying in *#06# in the keypad.

IMEI is useful for tracing stolen phones. A mobile network can block the use of a stolen phone by blocking the IMEI in the network. A phone with a blocked IMEI will not work even with a different SIM card on that network. IMEI is a good security feature, which is associated with every phone call taken or received even with different SIM cards.

Prepaid and post-paid phone payments

Prepaid

The cost for phone calls is paid in advance in a prepaid mobile connection. A prepaid user can make calls till he/she runs out of credit for the charged amount. Prepaid users do not receive monthly phone bills. Most prepaid phone connections do not incur a monthly phone rental. Prepaid phone credit limit can be increased (also called recharge, top-up) by various methods. It can be done using credit cards, over the phone, through Internet, SMS or at a retail outlet that can facilitate recharges for cash payment. Usually prepaid accounts have an expiry period and have to be recharged before date of expiry to keep the account active. Some premium and value added services may be restricted on prepaid phone accounts. Prepaid accounts are very popular among low budget phone users.

Post-paid

Unlike prepaid phone accounts, post-paid account holders receive a regular phone bill and do not have to pay in advance for the phone calls. Post-paid phone accounts usually incur a monthly rental or minimum charge.

112 Emergency number

112 is the universal emergency help access number for GSM and 3G mobile phones. In some occasions it is possible to dial this number even without a SIM card, when the phone keypad is locked or when there is no credit in a prepaid account. Calls to this emergency number 112 can be dialled from any country when network coverage is available and are connected to the emergency service operator or to the nearest police station. Misuse of this number can incur penalties as calls and their originating phone details are monitored and recorded.

Roaming

In mobile communication roaming refers to the use of a mobile phone in a mobile network (usually in a foreign country) which is different from the users own home network. The same mobile number is used to send and receive calls and SMS while roaming in the foreign network.

In most countries and networks roaming is automatically enabled when the phone is switched on in a visitor network area. In some networks roaming facility has to be activated on request and a payment or deposit may be required. The home network also should have a roaming agreement in place with the visitor network for proper roaming operation.

The phone automatically picks the roaming network when the phone is switched on in a visitor network area. The phone can be set to scan available networks and choose a network manually as well.

The mobile phone used in the visitor network should be a compatible with that mobile network for roaming operation. For example in USA GSM phones operate in 1900 MHz band and to roam in a USA GSM network a mobile phone that supports GSM1900 band should be used. Alternatively if the network uses a different type mobile phone the home network SIM card can be inserted to a compatible phone and used.

When a subscriber with a roaming mobile phone travels abroad all calls directed to his/her home mobile phone are routed to his/her phone in the foreign country. The roaming phone owner has to bear the international call cost for routing calls from his/ her usual home network to foreign network as a roaming service. To prevent this, international incoming calls can be blocked if required. The originator of the call to a roaming phone will only pay the ordinary call charges.

Roaming charges are usually higher than normal mobile phone charges and can lead to high phone bills specially if using wireless data access. To reduce costs, in some networks it is possible to enable only SMS roaming without enabling roaming for voice calls when travelling abroad.

Mobile Phone Frequency Bands

Mobile phone networks operate on different frequency bands depending on the country and the network. A single frequency band phone can operate in only one frequency band, dual band phone on two frequency bands, tri band phone on three frequency bands and a quad band phone in four frequency bands. Ability to operate in more frequency bands makes it possible for the phone to be used in different countries and networks.

In Europe and most other countries GSM phone networks operate on 900 MHz Band. In addition to this 1800MHz band is also used in urban areas. 850 & 1900 MHz bands are used for GSM is USA.

3G UMTS networks mostly use 2100MHz band. Some networks also use 850MHz Band for 3G (USA and Australia). 900 MHz band is also used for 3G in some countries.

LTE networks already deployed use different frequency bands varying from 700, 800, 850, 1700, 1800, 1900, 2100, 2300, 2600 MHz bands depending on the country and mobile network operator.

Mobile phone modes

In cellular mobile phones, mode refers to the type of transmission technology used; it can be GSM, CDMA, AMPS, D-AMPS, 3G, and LTE etc. Multi-mode phones support more than one mode of operation. Most 3G phones also support GSM or CDMA services and automatically switch to GSM/CDMA if 3G coverage is not available. Most satellite phones also have GSM/CDMA mode built in. Multimode phones are specially useful for roaming service when travelling to different countries where different types of phone networks are available.

Bluetooth

Bluetooth is a feature that allows a Bluetooth enabled device to connect wirelessly to another Bluetooth enabled device at close range. A Bluetooth enabled phone can connect with another Bluetooth enabled phone or a Bluetooth enabled computer to exchange information. Bluetooth is explained in detail in another chapter.

MMS

MMS is the abbreviation for Multimedia Messaging Service. MMS can be used to send messages that contain text, pictures audio and video. A MMS enabled phone can send an MMS to another MMS enabled phone using wireless mobile data connectivity.

Voice Mail

Voice Mail service enables a phone user to receive recorded audio messages from callers when the phone is switched off or not in a coverage area. When voice mail is enabled callers to a mobile phone have the option of leaving a recorded voice message when the phone is not reachable. Messages are stored in a server in the mobile network and can be retrieved later. When a voice message is recorded, the phone owner usually will receive an SMS when the phone is activated notifying about the voice message.

Missed call alert

Missed call alert feature enables a phone owner to know about attempted calls to his/her mobile while the phone was switched off or not in a mobile coverage area. When the phone is switched on, an SMS is sent notifying the date, time and originating number of the attempted call.

Call Forward

Call forward feature enables calls to a mobile phone to be diverted to another phone number. Different conditions can be set for diversion. Calls can be diverted if the phone is switched off, not answered after a given number of rings, without any condition (all calls diverted) or if the phone is busy. Usually the charge for the diverted call has to be borne by the mobile phone owner.

Call Waiting

Call waiting feature allows a mobile phone to get an indication of another incoming call while engaged in a phone call. The first phone call can then be put on hold to answer the second call.

Wi-Fi

Most new mobile phones have the Wi-Fi facility which enables a mobile phone to connect to a Wi-Fi network. Internet access is possible through the Wi-Fi network from the mobile phone with Wi-Fi capability.

Camera/Video Camera/Audio Recording/MP3/Music player

Most new mobile phones are equipped with camera, video camera and audio recording and playback features.

Video Calling

Video calling feature allows two people having a conversation using 3G Mobile phones with video calling feature within a 3G coverage zone to see each other's picture in video format on the phone screen.

Speakerphone

Speakerphone function allows the voice to be heard out loud in a call. The phone does not need to be held to the ear when using speaker phone function and even more than one person can take part in the call through the phone at one end.

GPS

Most new smart phones have built in GPS capability to get location information. A-GPS or assisted GPS is available in mobile phones use the mobile phone network signal in addition to the GPS satellite signal to determine its location and for navigation. Use of A-GPS function may incur additional data usage or surcharges.

Internet access

Mobile phones equipped with any of GPRS/EDGE/HSPA or LTE features can be used to access internet on the phone.

Tethering

The use of the mobile phone as a modem to connect to the internet from a computer is termed as tethering. In this instance the mobile phone will use the wireless mobile phone network data service to access internet.

Mobile Hotspot

In mobile phones the hotspot feature enables many other devices to connect to mobile phone using Wi-Fi and access internet using the mobile phone as a modem.

Caller ID

Caller ID or CLI facility enable a mobile phone user to see on screen the phone number of an incoming call. If the number is already saved in the mobile phone contact list the name of the person calling will be displayed on phone.

Memory Cards

Most new phones have the ability to install removable flash memory cards of different capacities which can be used to store information in the form of media files, photos, documents, mp3 files, songs etc.

FM Radio

FM Radio function enables to listen to FM Radio channels from a mobile phone. In most instances it is required to connect the mic/headphone cable to use this feature since it acts as an aerial/antenna.

FM transmitter

Some mobile phones have the FM transmitter function. This enables to play audio files in the phone through a car radio while travelling or through any FM radio in the vicinity.

HD (High Definition) Voice

HD Voice provides clearer crisper voice on mobile phone calls suppressing background noise such as traffic. HD Voice uses Wide Band Adaptive Multi-Rate (WB-AMR) coding which uses a wider audio frequency bandwidth that makes voice sharper. In order to make HD voice calls both originating and receiving mobile phones should have HD voice compatibility and the mobile phone network should support HD Voice feature.

NFC (Near Field Communication)

NFC is a method for establishing radio communication between two devices by touching them together or bringing them in to close proximity to each other, typically to less than about 10cm. It can be used to make contactless transactions, exchange digital content and set up connections between devices with a touch. NFC enabled mobile phones can be used to make payments by touching or waving it at NFC enabled payment terminals. In addition there will be many other applications that will be possible using NFC enabled mobile phones in future such as sharing information between phones, making payments for use of public transport, movie tickets etc.

Push to Talk

Push to Talk (PTT) over cellular is a system where a mobile phone user can talk to another one or more mobile phone users in a group by pressing a button in the phone similar to talking over a walkie talkie. When someone uses PTT and speaks, all the members of the group hear the message from the speaker of their mobile phone. To use PTT feature, the mobile phone network should support this feature and the users should have subscribed to the PTT service. In addition the mobile phones used also should have PTT facility.

In normal phone communication a connection is established for the whole duration of the call and it is a full duplex conversation with both parties able to talk at the same time. When using PTT, only one person can talk (half duplex communication) to his group at a time by pressing the PTT button in the mobile phone. Connection is active only while the PTT button is pressed. A group may consist of two or more number of persons and the group members have to be prescribed beforehand.

This feature is specially useful in a commercial environment where members of a workgroup can communicate with each other at the press of a button in short bursts of messages without having to establish a dialled connection.

(A service offering similar features is Trunked Radio service which is explained in a separate chapter).

Smart Phones

Most new mobile phones sold today fall in to the category of Smart phones. Smart phones are built on a mobile computing platform and popular mobile phone operating systems are Apple iOS, Android and Windows Phone. Smart phones have touch screens, Wi-Fi and Wireless Mobile data access, GPS, and many other features. Hundreds of third party applications can be downloaded for these platforms from the application stores for these operating platforms, some of them free of charge. These applications (popularly known as apps) are software designed to run on a mobile device and add some form of additional functionality to the mobile phone. These applications range from games, weather information, organisers, navigation aids, music applications, social networking, travel, sports, browse tools and hundreds of others types.

7

Wireless Mobile Data Services

Mobile phones, in addition to voice communication, provide data communication facilities in the form of SMS, MMS and internet access using a number of methods like GPRS, EDGE, and HSPA etc. Wireless mobile data services can be used to access internet on a mobile phone or on a computer using the mobile phone as a modem (tethering). Mobile wireless services can also be used to access internet on a computer without the use of a mobile phone by using a USB modem (dongle) attached to a computer.

SMS

SMS is the abbreviation for Short Message Service. SMS is a text message service where short text messages can be sent between Cellular Mobile Phones. SMS originated with the GSM mobile system and is also available with 3G phones and satellite phones. A text message typed on the mobile phone using the keypad can be sent to another mobile phone.

The size of a text message is usually limited to 160 characters. But most phones allow longer messages by breaking the message in to several messages and sending them as separate text messages and combining them at the receiving end.

SMS has several advantages. They are discreet, can be used by hearing impaired people, and can also be sent to multiple recipients. It is a cheap method to send across messages to phones even overseas.

In a situation where a text message is directed to a switched off mobile phone, the message is stored and is delivered by the network to the recipients when the phone is switched on. It is also possible to get a delivery confirmation report back by the originating mobile phone to confirm the text message has been delivered and time of delivery by enabling this facility in the sending mobile phone. The messages received are stored in the SIM card or mobile phone until they are deleted.

Text messages can be sent in languages and script other than English as well.

MMS

MMS is the abbreviation for Multimedia Messaging Services. MMS can be used to send multimedia content such as pictures, ringtones, audio & video files and animations. MMS compatible mobile phones should be used to send and receive MMS. Photographs, audio and video taken from camera equipped phones can directly be sent from the phone using MMS.

Circuit Switching and Packet switching

Having a knowledge about Circuit switching and Packet switching will help to understand about mobile data services. In circuit switching an unbroken connection is made between the connected parties throughout the duration of the call. Landline telephone calls and voice phone calls on mobile phones are circuit switched calls. Circuit switched calls are usually charged for the duration of the connection.

In packet switching digital data are compiled in to data packets and sent. The connection is active in bursts of time when there are packets to send across and then the connection becomes inactive. All high speed data services such as GPRS, EDGE, HSPA, use packet switching technology to send data. The charging for packet switched data is not done on the duration of connection but on the amount of data sent across. Amount of data transferred is measured in KiloBytes (KB) and MegaBytes (MB) and are charged according to the amount of data used in KB or MB. With these data services it is possible to have an always on data connection to the mobile phone network from a mobile phone.

Using mobile data services, internet browsing can be done on the mobile phone or a tablet. Alternatively using the mobile phone/tablet as a modem, internet can be accessed from a computer; the computer should be connected to the phone with a cable or with Bluetooth. Similarly the mobile phone/tablet can be made a Wi-Fi hotspot enabling many devices to connect to it using Wi-Fi and access internet through the phone/tablet using its wireless mobile connection. This is known as tethering.

It is also possible to access wireless broadband internet from a computer without using a mobile phone. Inserting a PCMCIA GPRS/EDGE/HSPA card to a Notebook computer to access wireless broadband is one method. A GPRS/HSPA/LTE USB wireless mobile broadband modem (similar in appearance to a USB memory stick) can also be connected to the computer to access internet without using a mobile phone.

Some of the latest notebook computers have built in mobile broadband access facilities.

In all above situations, a subscription to the mobile phone wireless data service is needed to access the broadband services

To use GPRS/EDGE/HSPA/LTE mobile data services, the phone or device, the phone network and the connected radio base station should support the relevant data service. In mobile phone networks all radio base stations may not be equipped to support these services. GPRS and EDGE connections are provided on GSM networks while HSDPA is a 3G network feature.

Adaptive Modulation

Adaptive modulation techniques are used in some mobile phone data transfer modes. In Adaptive modulation and coding, the mobile network figures out how good the connection is and changes the way it sends the data. A faster data rate is achieved when a good connection is available. Users near a mobile radio base station will get data more quickly than users far away from a radio base station where reception is poor. The data transfer speed will also become slower when the network traffic is high.

30

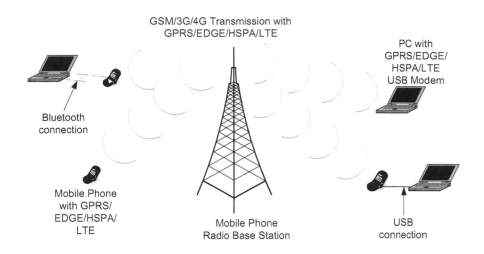

GPRS

GPRS is the abbreviation for General Packet Radio Service. It is a mobile phone data service of 2nd generation system GSM.

GPRS data is sent as packets of data and is a best effort service which means that data packets are sent when free channel time slots are available.

Actual GPRS data speed is dependent on the handset, available GPRS coverage and the amount of mobiles in use within the area. The theoretical maximum achievable GPRS data speed is 115kbps and the typical speeds are 20-40kbps.

There are three different types of GPRS mobile phones.

- Class A – GPRS and GSM (voice, SMS) services can both be operated at same time.

- Class B – Only one service, GPRS or GSM call is possible at a time. If GPRS services are active, during a voice call it is suspended and then automatically activated at the end of the call

- Class C – Only one service, GPRS or GSM is possible at a time and has to be switched manually.

31

EDGE

EDGE is the abbreviation for Enhanced Data rates for GSM Evolution

EDGE, which is an add on service to the GPRS service, is a method to increase data transfer rate on GSM phones using new modulation and coding schemes. EDGE can transfer data three times higher than GPRS. EDGE uses GMSK modulation with 8-PSK (8-Phase Shift Keying). Data transfer speed in EDGE can go up to 384kbps when all 8 timeslots in a channel are used.

HSPA

HSPA (High Speed Packet Access) is used to denote both HSDPA (High Speed Downlink Packet Access) and HSUPA (High Speed Uplink Packet Access). In HSDPA, more emphasis is placed on the data transfer rate in the downlink direction (radio base station to mobile phone) which is faster. The aim of HSUPA is to increase the uplink (mobile phone/device to radio base station) data transfer speed.

Adaptive modulation and coding are used to improve the data transfer rates in HSPA.

HSPA is provided on 3G networks and a 3G mobile phone or device with HSPA capability is needed to access these services. 3G service subscription, being within a 3G network coverage area and the connected radio base station having HSDPA/HSPA facility are other requirements to access these services.

HSDPA

In HSDPA, QPSK and 16 QAM modulation techniques are used. A theoretical maximum download speed of 14.4Mbps is possible with HSDPA but in reality the maximum data rates achievable is less than about 2-3 Mbps. Peak upload speed achievable is 384kbps.

HSDPA services can be usually used even while on a voice call.

HSUPA

HSUPA can provide an increased uplink data speed than HSDPA with peak uplink speeds of up to 5.76Mbps.

HSPA+

Also known as HSPA evolved. HSPA+ latest release is capable of providing up to a theoretical peak data speed of 42Mbps in the downlink and up to 11Mbps in the uplink. 64QAM data modulation for downlink and 16QAM modulation for uplink are used to achieve these higher speeds.

Dual Carrier/Channel/Cell HSPA (DC-HSPA) technology further increases data transfer speeds up to a theoretical maximum data speed of 84Mbps by utilising two parallel carriers/channels at a time.

LTE

Long Term Evolution is the technology which is evolved from 3G UMTS networks. Peak data transfer speeds of 100Mbps and beyond are expected to be achieved depending on network conditions, frequency and spectrum bandwidth used in LTE.

LTE Advanced

LTE Advanced is the potential technology for true 4th Generation (4G) services. Peak data transfer speeds of up to 1Gbps are expected in LTE Advanced.

Summary of Mobile Network Data services

Technology	Type of Radio Modulation	Description	Maximum speed	Typical speed - Downlink	Typical speed - uplink
GPRS	TDMA	General Packet Radio Service	115 kbps	20-40 kbps	20-40 kbps
EDGE	TDMA	Enhanced Data Rate for GSM Evolution (Enhancement to GPRS)	484 kbps	70-130 kbps	70-130 kbps
3G UMTS (without enhanced data services)	CDMA	3G Universal Mobile Telecommunication System	384 kbps (2 Mbps specified)	200-300 kbps	200-300 kbps
3G UMTS with HSDPA	CDMA	High Speed Downlink Packet Access	14.4 Mbps	0.5 - 4 Mbps	0.5-2 Mbps
3G with HSUPA (HSPA)	CDMA	High Speed Uplink Packet Access	14.4 Mbps	0.5-4 Mbps	0.5-3 Mbps
3G with HSPA+ (Release 8)	CDMA	Evolution of HSPA	42 Mbps	Above 5 Mbps	Above 3 Mbps
LTE	OFDMA	Long Term Evolution	326 Mbps	10 Mbps	5 Mbps
LTE Advanced	OFDMA	Long Term Evolution Advanced	1 Gbps		

Data transfer speeds indicated above are only approximate values. The data transfer speeds depend on many conditions such as traffic loading on network, conditions of the connection, distance to radio base station etc. The peak data transfer speeds indicated are rarely achieved in real conditions.

SMS Roaming and Mobile Data Roaming

Similar to Roaming service in voice, SMS and Data Roaming also can be used while roaming in a foreign network. The data connection type (GPRS, EDGE, HSDPA, HSUPA) will depend on the mobile handset type, roaming network radio frequency, features provided and the roaming agreement between the operators. Data roaming charges can be quite high and care must be taken in using as it can result in very high roaming charges. SMS only roaming is a cost effective alternative to avoid high data and voice roaming charges when travelling abroad.

Wi-Fi on Mobile

Most new mobile phones have Wi-Fi facility built in. Wi-Fi enabled mobile phones can connect to Wi-Fi networks wherever Wi-Fi coverage is available and the phone can be used to access internet or use any application in the mobile phone that requires internet access. Secure Wi-Fi networks require a password to connect to the network.

8

Femtocell

Femtocells are miniature low power cellular radio base stations that are placed inside houses and business premises. Femtocells are also known as Small Cells. Femtocells use an integrated antenna to transmit and receive low power mobile phone signals within a home or office. It is similar in operation to a Wi-Fi access point but is specially designed for cellular mobile phones.

Femtocells are connected to a DSL or cable modem. Femtocells communicate with the mobile phone network through the DSL or cable based broadband internet connection.

Femtocells are used to improve indoor coverage of mobile phone network signals. Mobile phone users can have the option of having their mobile phones connected to the mobile phone network when they are in their homes through a femtocell. A limited number of mobile phones can connect to a femtocell. Advantages of being connected to the mobile phone network through femtocells include better signal quality and less battery usage in mobile phone since the mobile phone will operate with lower transmit power. Mobile phone service providers may offer lower tariffs when using a femtocell to connect to the phone network. Improved signal strengths and quality can make mobile data transfer speeds higher when connected through a femtocell.

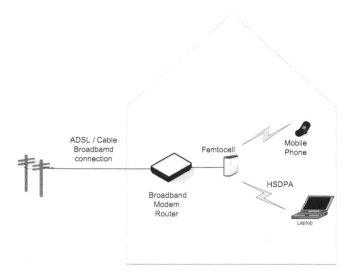

Femtocells offer seamless handover with outdoor mobile phone radio base stations. When a mobile phone user is on a call inside a house through a femtocell and moves out and travels while still engaged in the call, the call is handed over to the outdoor mobile phone network radio base station.

Femtocell Customer Premises Units

Use of femtocells can result in an increase in call/data capacity in a mobile network. Use of femtocells can offload mobile voice and data traffic from outdoor radio base station network and divert them through broadband internet connected femtocells to mobile phone users in homes and offices.

Femtocell products and standards are still in development stage. Already few mobile phone network operators have started deploying femtocells in their customer premises.

9

CDMA

CDMA is the abbreviation for Code Division Multiple Access. It is a wireless transmission technology used in CDMA Mobile phones and 3G Mobile phones.

CDMA was originally developed over 30 years ago for use in military communications It is difficult to intercept and decode or jam CDMA transmissions. CDMA employs spread spectrum technology with a special coding scheme.

In CDMA, each conversation is digitised and encoded with a unique code and transmitted over a wide radio frequency band. The radio frequency band is shared by other users also and carries hundreds of other conversations each of which is encoded with its own unique code. At the receiving device, by applying the unique code to the received signal the original speech signal is extracted.

Most 2G cellular mobile phone systems like GSM use different sets of frequency channels in neighbouring cells to avoid interference. In CDMA based cellular mobile systems like CDMAOne and UMTS (3G), same frequency channels are reused in neighbouring cells. Although signals from neighbouring cells appear as interference, by applying the correct code in the demodulation and signal extraction process, the required signal can be extracted.

CDMA Systems

CDMAOne

CDMAOne is the 2nd Generation CDMA based digital Mobile phone system which is in use in over 30 countries including Korea, USA and Japan. This system is also known as IS-95. IS-95A provided voice and data speeds up to 14.4kbps. IS-95B which was the upgraded system from IS-95A provided data transfer speeds of up to 64kbps. IS-95 system is also used as a fixed wireless local loop telephone system in many countries

CDMA2000 1X

CDMA2000 1X was developed from IS-95 and provides about twice data and voice capacity of IS-95. CDMA2000 1X is considered as a 3rd generation technology.

CDMA2000 1X supports data transfer speeds up to 144kbps.

CDMA2000 1xEV-DO

An improved version of CDMA2000 technology called CDMA2000 1xEV-DO is used for wireless broadband access. (1X EV-DO stands for 1 carrier Evolution-Data Optimised). Maximum data rates of up to 3.1Mbps downlink and 1.8Mbps uplink are supported with this technology. In this system data and voice cannot be used in the same channel.

CDMA2000 1xEV-DV

CDMA2000 1xEV-DV is another version (1X EV-DV stand for 1 Carrier Evolution – Data and Voice) which can support downlink data rates of up to 3.1 Mbps and Up Link rates of up to 1.8 Mbps. It also supports concurrent operation of voice users, and high speed data users within the same radio channel.

3G (UMTS) Mobile

3G UMTS Mobile phones use a CDMA system known as WCDMA (Wideband CDMA)

Soft Handover

Handover in non CDMA based cellular mobile phone systems used hard handover when a phone user, while engaged in a call, moved from one radio base station area to another. When the signal from a radio station becomes weaker while on a call and moving away and signals from another radio base station becomes stronger, a new channel is assigned for the call from the radio base station where the signals are becoming stronger. The connection from the earlier base station is broken and the call is rerouted through the newly assigned channel of the new radio base station instantly without interrupting the call. This is known as hard handover which is used in non CDMA based cellular mobile phone networks.

In handover situations in CDMA based mobile networks the mobile phone connects to a new radio base station when a phone user while engaged in a call is goes away from the already connected radio base station and nears another radio base station. The initial connection is not immediately terminated here. For some time the phone communicates through both radio base stations before the connection through the earlier radio base station is eventually dropped. Then the communication takes place only through the newly connected radio base station. This is known as soft handover and is used in CDMA based mobile phone networks like CDMAOne and 3G (UMTS).

CDMA Fixed Wireless phones

CDMA fixed Wireless local loop telephones work similar to the mobile phone network operations. The phones communicate using radio signals with the nearest CDMA radio base station. CDMA phone should be within the signal coverage area of the particular network service provider Radio base station for it to function. However unlike in cellular mobile, mobility and handover is limited.

A CDMA fixed wireless Phone

CDMA Wireless local loop technology is very much useful for providing wireless telephone connections in areas where provision of land line telephone connections are not viable due to non-availability of wire line telephone network or unavailability of loops. CDMA fixed wireless phones are mostly used in developing countries that do not have a developed wire line phone network infrastructure.

10

WiMax

WiMax is the abbreviation for Worldwide Inter-operability for Microwave Access. WiMax is a wireless data transmission technology. It uses wireless transmission for broadband data transmission and internet access. Its configuration is similar to a mobile phone networks and comprise of WiMax radio base stations providing network coverage to fixed or mobile WiMax receiver devices. End user devices can be notebook computers with inbuilt WiMax access capability or network access devices connected to computers. WiMax is an attractive technology for last mile broadband access where it is difficult or not viable to lay cables to customer premises. WiMax is a certification trade term given to equipment compliant with IEEE 802.16 specifications.

WiMax technology can also be used in point to multipoint network or in a mesh (ad hoc) network where stations establish connections directly with other stations.

WiMax is based on the IEEE 802.16 standard.

There are three subdivisions of this standard.

• IEEE 802.16d (also known as 802.16-2004) for fixed broadband access.

• IEEE 802.16e (also known as 802.16-2005, WiMax release 1) for Mobile broadband access.

• IEEE 802.16m (also known as WiMax release 2) which is an improved version of WiMax release 1 and backward compatible with release 1)

(Original IEEE 802.16 or 802.16a standard specified 10-66GHz range frequencies while later IEEE 802.16d standard added the 2-11 GHz range.)

WiMax uses OFDM (Orthogonal Frequency Division Multiplexing) technology for radio frequency modulation. WiMax can operate on frequencies between 2-11 GHz depending on the country and the network operator. For mobile WiMax 2-5GHz band frequencies are used.

WiMax can support data transfer speeds of up to 70Mbps; the higher speeds are achieved at short range using maximum channel size and maximum modulation. Mobile WiMax data rates can typically be up to about 15Mbps and the cell radius about 2-4kms.

There are many fixed WiMax service networks in operation throughout the world. Since the introduction of LTE services the demand for WiMax services are declining.

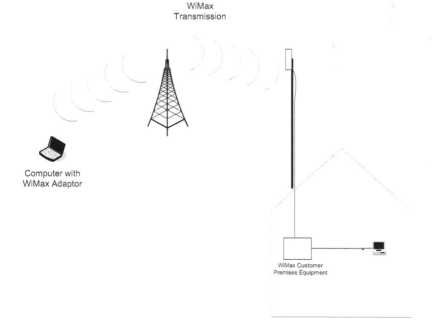

WiMax
Transmission

Computer with
WiMax Adaptor

WiMax Customer
Premises Equipment

Adaptive Modulation

WiMax supports a number of adaptive data modulation and FEC (forward error correcting) schemes which can change for individual users depending on radio channel conditions and subscription level. BPSK, QPSK, 16QAM and 64QAM modulation methods are used in WiMax.

WiMax Antennas

WiMax uses improved antenna technology such as MIMO (Multiple Input Multiple Output) technology which include Space-Time Diversity Coding, Spatial Multiplexing, and Adaptive Antennas Array Systems using beam forming or beam tracking technology. Both indoor and outdoor units are available for fixed WiMax access. Indoor units are suitable when signals from WiMax radio base stations are strong; if signals are weak outdoor unit is more suitable. An outdoor WiMax unit is the size of a notebook and is mounted on a pole outside. For fixed WiMax access having an outdoor WiMax antenna unit allows the subscriber unit to be further away from a base station than an indoor unit.

Bandwidth

The bandwidth of a WiMax channel can range from 1.25 – 20 MHz depending on the availability of the radio frequency spectrum and the allocation to the WiMax service provider. Both TDD (Time Division Duplex) and FDD (Frequency Division Duplex) formats are supported where either a single channel or a pair of channels for uplink and downlink can be used. Mobile WiMax uses only the TDD scheme.

WiMax networks are based on an all IP (Internet Protocol) architecture where end to end services are delivered on IP based protocols. Voice phone calls on WiMax use Voice over Internet Protocol (VoIP) technology.

WiMax Radio Frequencies

WiMax uses 2-11GHz range for consumer applications, 2.3, 2.5 and 3.5 GHz bands are the most widely used. WiMax mobile systems are designed to be operated on bands below 4GHz. 700MHz band is also expected to be available for WiMax in future after it becomes vacant with the analogue television switchover to digital television. WiMax system can operate on frequencies of 11- 66GHz also; these higher frequencies are used mostly for point to point links.

Wi-Fi vs. WiMax comparison

Wi-Fi is based on IEEE 802.11 standard and is designed typically for indoor wireless area network services. The range of an access point in a Wi-Fi network is about a maximum of 100m. Data throughput is 300 Mbps maximum. Wi-Fi uses 2.4 and 5 GHz bands for operation.

WiMax offers higher data speeds over longer rangers in outdoor environments. WiMax coverage range can go up to a maximum of 50km depending on the frequency, location and height of radio base station antenna. Typical range is from 3-10km for non-line of sight connectivity. WiMax data speeds can go up to a maximum of 70Mbps.

WiBro

WiBro is a wireless mobile data access system similar to WiMax developed in Korea. The developers of WiBro have agreed to harmonise the technology with WiMax in future.

WiMax Release 2

A new version of mobile WiMax 802.16m also known as WiMax release 2 that would increase data transfer speeds up to 1Gbps while maintaining backward compatibility with existing WiMax products will be released in future.

WiMax release 2 can be a competing technology along with LTE-Advanced (Long Term Evolution–Advanced) for 4G standard communication systems.

There are over 500 WiMax networks in operation in 140 countries.

11

Satellite Telephone Service

A satellite telephone is a mobile telephone that uses satellites for communication. Satellite phones directly communicate with an orbiting communications satellite and calls are connected to Public Telephone service through a ground based Gateway station. Gateway stations have large dish antennas and connect the satellite phone calls to wired or cellular mobile phone systems. Satellite phones which are also called Sat phones are useful in remote areas where cellular mobile phone coverage or any other communication facilities are not available. GMSS (Global Mobile Satellite Systems) is the term widely used for these types of systems.

Satellite phones are usually much more expensive than cellular mobile phones. Their usage charges are also more expensive. Satellite phones provide voice telephone and data services and are used in international news reporting, remote expeditions, emergency communications, rescue and disaster relief operations, on board ships, maritime & aviation, oil & gas rigs etc.

Satellite telephones are specially useful in emergency communications and disaster relief operations where normal communication services have been destroyed or are overloaded.

The size of the satellite telephone depends on the type of satellite phone system. Some systems have handheld units small as normal cellular mobile phones, some are bulkier handsets while some are larger units which are the size of a notebook computer or briefcase.

46

Satellite phones need to have clear view of the sky and line of sight to the satellite for them to work properly. They may not work indoors or when their signal path to satellite is blocked by some obstruction. To operate from a confined or indoor location an external outdoor antenna may be required.

Calls made to satellite telephones from landline or mobile phones can be quite expensive, even higher than the cost for an international call.

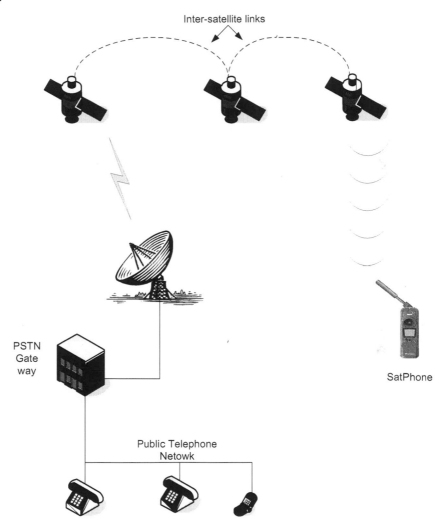

There are two types of Satellite telephone services providing services by Geostationary (GEO) satellites and LEO satellites as explained below.

Geostationary (GEO) satellites

These systems can provide wide coverage with less number of satellites. One satellite can provide coverage to about one third of the world. 3-4 satellites can provide worldwide coverage. Satellites are placed in geosynchronous orbit 35,786km above earth. These satellites orbit the earth at the same rate as the earth's rotations and appear stationary relative to earth. Due to the large distance to satellites it takes nearly a quarter second for signals to travel from earth to satellite and return and this delay is noticeable in telephone conversations. The large distance to satellite also means more losses to signals, higher power and larger antennas are required for these systems than for LEO satellite type systems.

Thuraya and Inmarsat satellite phone systems use geostationary satellites.

LEO satellites

LEO (Low Earth Orbit Satellites) use satellites positioned 500-1500km above earth. LEO satellites orbit the earth at high speeds (approximately 7km/sec) and takes 70-90 minutes to complete one orbit.

Each satellite provides coverage to only a certain area at a time and many satellites are required to provide worldwide coverage. Some of the LEO type satellite phone systems have inter satellite links with on-board processing that allow communications between neighbouring satellites while others have satellites which only relay the signal back to a ground based Gateway. As LEO satellites constantly travel in the sky, a satellite will be visible from one position on earth between 7-20 minutes at a time depending on satellite altitude and user position. Satellite phones for these types of systems are much more compact than in GEO type satellite systems because they require less power to transmit and have smaller antennas due to lower distance to satellites.

Iridium and Globalstar satellite phone systems use LEO satellites.

Dual Mode Cellular/Satellite Phone

Dual Mode Satellite/Cellular phones enable the user to place a call using either the cellular mode or the satellite mode. In the cellular mode the call is connected to the terrestrial cellular mobile network while in the satellite mode the call is connected via a satellite. Most of the satellite phones in the market are Dual Mode phones as they enable the user to place and receive calls using cellular networks where call charges are cheaper and to use the satellite mode only when cellular mobile coverage is unavailable.

International Dialling Access Code

ITU (International Telecommunication Union) has allocated country code +881 followed by a single digit for each operator for Global Mobile satellite services. However Inmarsat satellite phone service is an exception which was earlier allocated country codes +870 to +874. Some satellite services launched later use the +882 shared country code followed by two digit operator code.

(+ sign stands for international direct dial access code, which usually is 00 in many countries)

There are several different systems of satellite phones in operation. Some of them cover the whole world while some cover only certain regions of the world.

Kyocera Iridium satellite phones

49

Iridium Network

The Iridium satellite phone network was originally designed to have 77 satellites. It was named after the chemical element with atomic number 77 which is Iridium. Although the system was later launched with 66 satellites, the Iridium name was retained.

It is the largest commercial satellite constellation in the world. Iridium system facilitates worldwide voice and data communication using handheld Sat phones. It covers the whole earth including the oceans and Polar Regions. It is the only commercially available communication system that provides service throughout the earth. Due to security reasons Iridium services are restricted in some countries and regions. Some of the tsunami warning systems use Iridium phone system to send warning alerts.

Iridium system was spearheaded by Motorola company and commenced operations on Nov1 1998. US$5700 Million was spent on the system but after two years it became a financial failure.

The rapid advances and deployment of cellular mobile telephones which also had roaming services enabling to operate in many countries resulted in a low demand for satellite telephones since they were more expensive and had higher call charges. Iridium service went in to bankruptcy in August 1999. The satellites were about to be de-orbited and allowed to burn up in atmosphere when a group of investors bought the whole system for US$25 million which was less than two hundredth the amount it had cost. The services were revived by the new investors and the system is now back in operation. Iridium provides Fax, data, numeric and alphanumeric paging, voice, messaging (SMS), and prepaid services.

Iridium Satellites

The Iridium satellite constellation consist of 66 LEO (Low Earth Orbit) interlinked satellites, and several spare satellites. The satellites are in 6 orbital planes 780 km above earth, each plane having 11 satellites. Each satellite transmits at a power output of 400 watts. Iridium satellites circle the earth once every 100 minutes traveling at a rate of 16,832 miles per hour. Each satellite is cross-linked to four other satellites; two satellites in the same orbital plane and two in an adjacent plane.

Iridium Ground and Control Stations

The satellite network operations centre is located in Virginia USA with a backup centre in Arizona USA. Iridium commercial gateway station is in Arizona USA which provides interconnection between satellite network and public telephone network. The Gateway handles call set up procedures and interfaces the Iridium system with the Public Telecommunications Services. Calls from one Iridium phone to another Iridium phone are routed through inter-satellite links without going through the Gateway station.

Four tracking and control stations are positioned at strategic positions around the world.

Iridium satellite phones operate in the L band frequencies 1616 – 1626.5MHz.

The Iridium International Dialling access codes are +8816 and +8817, which is followed by an 8 digit Iridium phone number.

Motorola Sat Phones

Iridium Next

Iridium Next will be the next generation of satellites which will replace the current Iridium satellites. New satellites will provide additional features with improved data transmission rates and will be an end to end IP (Internet Protocol) based system. Iridium Next satellites are expected to be launched from 2015.

Globalstar

Globalstar satellite system consists of 40 LEO satellites and 4 spare satellites.

This system uses CDMA signalling and a call from a Globalstar sat phone may be connected with more than one satellite at a time. This path diversity in CDMA signalling ensures that even if the signal to one satellite gets interrupted the call is not interrupted due to soft handoff feature where other satellites can maintain transmission paths.

Globalstar covers 80% of the earth's surface that include over 120 countries. (Polar Regions are not covered). The satellites are not interlinked and do not communicate with each other. Therefore a satellite must have a Gateway station in view to provide service. Currently some parts of the world including India and a major part of Africa are not provided service by Globalstar system as there are no Gateways in those regions although satellite coverage may be available.

First Globalstar satellite was launched in Feb. 1998. In 1998 September, 12 Globalstar satellites were destroyed while being launched when the Ukraine made rocket crashed in Kazakhstan, Russia on its launch mission. In Feb. 2000 Globalstar commercial operations commenced. The satellites orbit at an altitude of 1414kms, orbit period of each satellite is 114 minutes. Satellites are in 8 planes, Satellite Transmitter Power is 380Watts Watts. Globalstar system has 24 Ground based Gateways.

In February 2002 like Iridium, Globalstar also faced bankruptcy. The US$ 4300 Million Globalstar system was bought for US$43Million by another company at one hundredth of the original cost.

Commencing from October 2010, a second generation satellite constellation of 48 satellites are being launched.

Globalstar Sat phone international access code is +8818, +8819, but most Globalstar Sat phones have been allocated USA phone numbers.

Globalstar Sat phones operate in 1.6GHz frequency L band for uplink (Sat phone to satellite) and 2.4GHz frequency S band for downlink (Satellite to Sat phone)

Asia Cellular Satellite (ACeS)

ACeS system uses two geostationary satellites, Garuda I and Garuda II. Three Gateways are in operation in Indonesia, Thailand and

Philippines. Coverage is limited to Asian Region. Handheld and Portable units are available. International Access Code +88220

Ericsson R-190 GSM/Satellite Dual Mode Phone for ACeS

MSat

MSat is a regional satellite phone system providing services to North America.

Inmarsat

Inmarsat was the World's first mobile satellite communications service provider.

Inmarsat started in 1979 as an Inter-Governmental Organisation and was earlier known as International Maritime Satellite Organisation. In 1999 Inmarsat was converted to a private company. Inmarsat services are available in 98% of areas in the world.

Inmarsat satellites

Inmarsat Satellites are of geostationary type and are placed 35,756 km above earth. Inmarsat system consists of a series of satellites. 4 Inmarsat I-2 second generation satellites were launched in early 1990s, 3 are still in operation, one has been decommissioned and is now parked in the higher Graveyard orbit safely out of way of working satellites. In 1996-1998 five Inmarsat I-3 satellites were launched and all are still in operation. Two next Generation satellites of Inmarsat called Inmarsat I-4 have also been launched and a third is to be launched in future. Inmarsat I-4 range of satellites provides superior services than the earlier satellites including broadband data services.

29 Gateways are in operation in 29 countries. Inmarsat Network operations centre is located in Inmarsat Head Quarters London.

Some of the Tsunami warning alert systems use Inmarsat system.

Inmarsat service is mostly available for portable units, which are the size of a notebook computer. Inmarsat recently introduced handheld satellite phones which communicate with Inmarsat-4 range of satellites

Inmarsat international Access Code is +870

Inmarsat portable satellite phone

Thuraya

First Thuraya satellite was launched in October 2000, the Second satellite in June 2003 and a third satellite were launched in January 2008.

Thuraya provides services to 140 countries in Europe, North, Central Africa and large parts of Southern Africa, the Middle East, Central and South Asia and Australia. Thuraya satellites are positioned in Geosynchronous Orbit, 35,786 km above the Earth. The Primary Gateway of Thuraya system which comprises the satellite control station and Public Telephone system Gateway station is located in Sharjah, United Arab Emirates.

ThurayaDSL is an internet connectivity service provided by Thuraya network where 144kbps data speeds can be obtained using a mobile terminal which is the size of a notebook computer.

Thuraya international access Code is +88216

Thuraya sat phones operate in 1.6 GHz (uplink) and 1.5 GHz (downlink) frequency bands.

Thuraya dual mode sat phone with GSM facility and built in GPS

Thuraya DSL portable unit for packet data connectivity up to 144kbps.

12

Bluetooth

Bluetooth is a short-range radio communication technology that enables wireless connectivity between electronic devices to communicate with each other. Bluetooth can connect and exchange data between Bluetooth enabled electronic devices such as mobile phones, mobile phone headsets, tablets, computers and peripherals. Bluetooth can be considered as a cable replacement method where devices can communicate at short ranges using radio waves without having to attach or connect using cables.

Bluetooth name

The word "Bluetooth" is named after the 10th century Danish King Harald Blatand. The name Blatand means Bluetooth after translation. King Blatand had played a prominent role in uniting Scandinavian Europe torn apart by wars. Bluetooth technology was first developed in Scandinavia and since it unites and connects different devices such as the cell phone, computers the name Bluetooth was chosen.

Bluetooth logo

Some examples of Bluetooth use

- A mobile phone connected wirelessly to a Bluetooth headset to enable a user to talk hands free. It is easier for the user to talk without the need to hold the phone to the ear or using a wired headphone/mic connection.

- Two mobile phones exchanging contact information using Bluetooth

- A computer sending music files/pictures to a mobile phone through a Bluetooth connection.

- A Bluetooth headset is connected wirelessly using Bluetooth to a PC used for a VoIP call using an application such as Skype, Yahoo, or MSN Messenger.

- A computer and a tablet or a mobile phone connected through Bluetooth can synchronise the contact/address book details in each other.

- Use of a mobile phone as a wireless modem with a Bluetooth enabled PC to access internet (this is known as tethering)

Bluetooth Mobile Headset

Bluetooth Wireless Connection

Mobile Phone

computer

Bluetooth Wireless Connection

Mobile Phone

USB Bluetooth dongles

Most new notebook computers now have built in Bluetooth capability. For computers without built in Bluetooth, an external USB Bluetooth dongle can be connected to transfer information from computer to another Bluetooth device.

A USB Bluetooth Dongle   **A Bluetooth mobile headset**

Bluetooth Radio connection

Bluetooth radio operates in the ISM (Industrial, Scientific and Medical) band at 2.4 – 2.485 GHz radio frequency. Bluetooth is a communication protocol and standard (IEEE 802.15.2) and consumes low power and transmits in low power. Bluetooth communication range is from 1-100 metres depending on the class of the Bluetooth device. Most widely used Bluetooth devices are Class 2 devices that operate within 10 metre range. Since Bluetooth uses radio waves that can travel through walls and non-metal barriers, connected devices do not have to be within line of sight, they could even be in two separate rooms as long as they are within the communication range.

Bluetooth uses frequency hopping spread spectrum (FHSS) technology. The frequency used changes at a fast pace and prevents unauthorised access.

Passkeys are used to ensure that only units authorised by their owners can communicate.

Bluetooth Device classes

Class	Maximum Permitted Transmit Power	Range (approximate)
Class 1	100 mW	100 m
Class 2	2.5 mW	10 m
Class 3	1 mW	1 m

Bluetooth Versions and Maximum Data Rates

Version 1.2   -   1 Mbps

Version 2.0+EDR   -   3 Mbps

Version 2.1+EDR   -   3 Mbps

Version 3   (HS)   -   24 Mbps

(EDR – Enhanced Data Rate,   HS – High Speed)

Later versions of Bluetooth devices are backward compatible with earlier version devices.

Connecting Bluetooth devices

A Bluetooth enabled device can usually connect and communicate with up to seven other Bluetooth devices at a time depending on the device capability.

Bluetooth devices usually broadcast their device name and type (printer, phone, camera etc.) continuously when switched on. Every Bluetooth device has a unique address (an identification number). A Bluetooth device user can enter a name for most Bluetooth devices which will be shown to other scanning Bluetooth devices. Bluetooth devices can set their discoverable state to ON or OFF. If the discoverable state is ON, the device name will be seen by other Bluetooth devices when they scan and search. When discoverability is OFF, the device will not appear when other devices scan and search for it. Undiscoverable devices can still communicate with other Bluetooth devices, but they must initiate all the communications themselves.

Bluetooth pairing

When two Bluetooth devices agree to communicate with each other, Bluetooth pairing occurs. When pairing is done the two devices are called a trusted pair. A trusted pair where authentication has been done once will not need to be authenticated on subsequent occasions and will be paired automatically using the existing link key.

Passkey (PIN)

Bluetooth devices have a passkey, which is a code that two devices agree to share in order to communicate. In some devices like Bluetooth mobile phone headsets which do not have any input method the passkey is fixed while in devices like computers and mobile phones the passkey can be changed and defined by the user. The passkey is used to prevent unauthorised access to Bluetooth devices, it can ensure that nobody can listen or intrude in to what a notebook computer sends to a mobile phone.

For pairing, both participants must agree on a passkey and enter it on device. An identical passkey must be entered on both devices. The passkey is an arbitrary code that is entered when requested; it is usually a 4 digit number. After the passkeys are compared and verified to be correct, paring occurs.

13

The Internet

Internet is a global collection of computer networks and servers which are interconnected. Internet carries a vast amount of information and is used by internet users to browse the World Wide Web, for media streaming, Voice over Internet Telephony (VOIP), Email etc.

To access Internet the following are needed.

- A device to display information, this can be a computer, a mobile phone, a tablet or similar device.

- A connection to an Internet Service Provider (ISP), this can be a modem with a dial up phone connection, ADSL or cable broadband connection, or a wireless mobile connection.

- A user account with an ISP

Fixed line and Mobile Telephone network operators that provide the connections are ISPs in most instances though there can be separate independent ISPs as well.

ISPs provide connection to a larger network which is part of internet. Connecting a computer to an ISP makes that computer to become part of internet.

Information available on internet are hosted on thousands of internet web servers of different computer networks. Private, public, business, academic, and government entities own these networks. Information on these web servers can be instantly accessed from anywhere in the world using an internet connection.

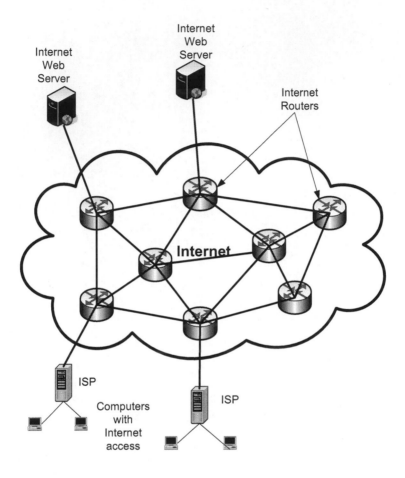

Some of the popular uses of Internet are given below.

WWW – World Wide Web

World Wide Web is a huge collection of information resources which are interlinked. To browse internet, an internet browser which is a software application such as Internet Explorer, Firefox or Chrome is used. Using an internet browser it is possible to navigate through internet web pages using hyperlinks which point to related information at other locations. To search for specific information the internet address should be known or a search engine like Google, Yahoo or Bing can be used. Entering keywords for the required information to the search engine will give thousands of pages of related information available on Internet. Information available could be in the form of text, graphics, pictures, audio, video, news etc.

Email

Electronic Mail which is known as Email is sent through Internet. To use email one should have an email address. Having an email account with an ISP is required to get an email address or alternatively a free web based email address can be easily registered using providers like Gmail, Hotmail, and Yahoo etc. In addition to messages, files in the form of documents, audio, video could also be attached and sent with emails.

Internet Telephony

Internet Telephony is a popular application using VOIP (Voice over Internet Protocol) to have voice conversations over the Internet. It is possible to have voice conversations free of charge between two persons connected to internet even in different countries using popular applications such as Skype, Yahoo, and Windows Live etc. If a web camera is available a video conversation could be carried out. It is also possible to make calls to normal telephones in different countries using these applications at a lower cost than that of an international telephone call by opening a paid account.

Remote Access

Remote Access allows an employee to connect to an office or commercial computer network that operates a VPN (Virtual Private Network) from a remote location through the Internet. This feature is useful for a worker to connect to his/her office network from a remote location or to work from home using Internet.

Streaming media

Audio and Video services like radio channels and television channels could be accessed over the internet. Thousands of radio and television channels from different countries are available for access over internet. YouTube is a popular site where Video clips could be uploaded by individual users which could be viewed worldwide by internet users.

14

Wi-Fi

Wi-Fi is a wireless networking technology that is used to interconnect computers and related equipment using radio waves rather than cables within a limited area.

Examples of Wi-Fi use are:

• A computer can send documents to printer for printing using Wi-Fi. The computer is not physically connected to printer with a cable. They are connected together using Wi-Fi, both computer and printer having Wi-Fi capability.

• Workers in an office are connected to the office WLAN (Wireless Local Area Network) through Wi-Fi connections. Each worker's computer is not physically connected using cables but by radio waves. One or more wireless Access Points (AP) which are connected to the office computer network and servers send and receive data from individual computers. For such a network wireless Access Points should be available. Computers should be Wi-Fi enabled and be within the range of an Access Point. Wi-Fi Networks in a building eliminate the need to use a cable network to connect computers to gain access to office network or access internet through a gateway. WLAN network give staff the freedom to move around and allow all the users to share network devices like printers and access internet.

The office WLAN network also makes it easy to add computers to the network. There is no need to install new cabling. For a computer which does not have built in Wi-Fi facility, adding a Wi-Fi card/USB Wi-Fi device and configuring it enables one to connect to the office network in minutes.

• A computer using Wi-Fi connection connected to a Wi-Fi enabled router can access the internet. The router can be a combined ADSL/Cable modem and Wi-Fi router unit that is connected to the public internet using the phone line or cable connection. The Wi-Fi router and computer are not physically connected using a cable; they are connected using Wi-Fi. Both computer and modem/router unit should have Wi-Fi capability.

• A notebook computer user (computer should have Wi-Fi capability) can access internet using Wi-Fi in a public wireless (Wi-Fi) hotspot where a Wi-Fi is available for public use. Airports, some cafes and hotels have public wireless hotspots. Access to Wi-Fi in public hotspots may be free of charge or in some places the user may have to make a payment.

• Two or more Wi-Fi enabled computers can connect with each other and transfer data with among them. The computers are connected using Wi-Fi and cables are not used. This type of connection is known as a peer to peer (P2P) connection.

Home Wireless Network

A home Wi-Fi network enables several PCs within a house to access each other's computers, send files to printers and share a single Internet connection.

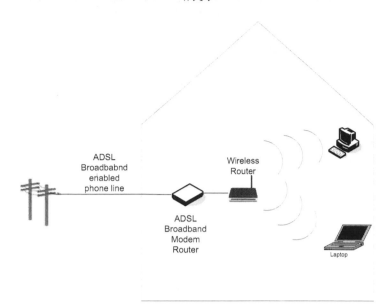

A Home Wireless Network

Wireless enabled devices.

To use Wi-Fi, the computer or related device should have Wireless capability.

A wireless network adaptor is required for a PC that needs to be connected to a wireless network. These are also called Wireless NICs – Network Interface cards. Wireless network adapters contain a radio transmitter and receiver (transceiver). Most new notebook PCs have built in wireless network adaptors.

It is possible to insert a Wi-Fi radio embedded in a simple PCMCIA (Personal Computer Memory Card International Association) card (known as a PC Card) into the Notebook PC expansion slot when Wi-Fi capability is not built in. When a desktop PC does not have built in Wi-Fi, an internal Wi-Fi network interface card (PCI card) can be installed to the PCI slot to use Wi-Fi. There are external USB Wi-Fi devices which look similar to USB memory sticks that can be used for Wireless access for computers without built in Wireless capability. Some printers and most new smart mobile phones also have built in Wireless capability.

A USB Wi-Fi adaptor

Peer to Peer Network (Ad-hoc mode)

A peer-to-peer network comprises several Wi-Fi equipped computers directly talking to each other without using an access point or gateway. This type of network is useful for transferring data between two or more computers or sharing an Internet connection among a few computers in a room.

Wireless Routers

Wireless Router is a wireless access point with the functionality of a Router. Wireless Routers combined with ADSL/Cable modems can be connected to an ADSL enabled phone line or cable connection and a home/office wireless network can be set up. Computers with Wi-Fi capability connected to the home wireless network can access internet.

A Wi-Fi router

IEEE 802.11

IEEE 802.11 is a set of international standards defined by the IEEE (Institute of Electrical and Electronics Engineers) for Wireless Local Area Networks.

The original 802.11 specification supported only 2Mbps of data and is no longer used. 802.11 was expanded to support higher data speed versions a, b, g and n.

802.11a

802.11a operates in 5GHz (5.15-5.35GHz) radio frequency band using 8 channels. Radio frequency modulation scheme used in 802.11a is OFDM (Orthogonal Frequency Division Multiplexing). It is less prone to interference than 802.11b. Maximum data speed defined is 54Mbps but practical achievable data speeds are less than 25Mbps. As the operational frequency is higher than 802.11b and 802.11g the operational range is less in 802.11a since the higher frequency signals have more difficulty in penetrating walls and other obstructions.

802.11b

802.11b operates in the 2.4GHz radio frequency range and can use 3 non overlapping channels. Bluetooth devices, some digital cordless phones and microwave ovens operate in the same frequency band and can sometime cause interference to Wi-Fi in this band. The maximum data transfer speed is 11Mbps, in practice it is about 4 - 6Mbps. The type of radio frequency modulation scheme used is DSSS (Direct Sequence Spread Spectrum). DSSS is less susceptible to radio noise and interference since it spreads a transmission signal over a broad band of radio frequencies

802.11g

802.11g is also known as Wireless–G. 802.11g uses the frequency band 2.4GHz which is the same as in 802.11b.

802.11g is backward compatible with 802.11b.

Peak Data speed is 54Mbps, but realistic data speed varies from 7-16 Mbps. 802.11g uses OFDM modulation. It is similarly susceptible to interference from other devices that share this band.

802.11n

This is the latest 802.11 specification. Maximum data rate is between100-600Mbps. This standard was designed to improve the data rate of 802.11g and use both 2.4 and 5 GHz frequency bands. A technology called MIMO (Multiple Input Multiple Output) using multiple antennas to send and receive parallel streams of data at one time is used to obtain higher data rates. Multiple antennas numbering up to 4 can be used in 802.11n. 802.11n uses OFDM as the radio frequency modulation scheme.

802.11n standard operate either in the 2.4GHz and 5GHz radio band, or both bands, offering backward compatibility with existing 802.11a/b/g equipment. To benefit from increased data rate capability of 802.11n, both connected network interface equipment should be of 802.11n standard.

The speed of 802.11n depends on the standard or optional features being used. Standard features use two 20MHz channels while optional features use two 40MHz channels, a 40MHz channel providing up to double the date rate of a 20MHz channel. Use of four 40MHz channels enables peak data speeds up to 600Mbps.

802.11n wireless router

Compatibility of 802.11 a, b, g and n versions

As 802.11a and 802.11b utilize different frequencies, the two technologies are incompatible with each other.

Some vendors offer hybrid (dual band) **802.11a/b** network equipment. These products implement the two standards side by side which means each connected device have to use one or the other of the two standards.

802.11g is backwards compatible with 802.11b and use the same frequency, meaning that 802.11g equipment will work with 802.11b equipment.

In simple terms

802.11a is compatible with 802.11a and 802.11n.

802.11b is compatible with 802.11b, 802.11g, and 802.11n.

802.11g is compatible with 802.11b, 802.11g, and 802.11n.

802.11n is compatible with 802.11a, 802.11b, 802.11g, and 802.11n.

When equipment interconnect in different standards the data transfer speed will be the lower value speed of the two standards. For example when 802.11b equipment and 802.11g equipment interconnect, the maximum data speed would be of the lower value of the two specifications which is 802.11b (11MHz).

Wi-Fi Data transfer speed

The data transfer speed depend on number of factors which include distance between the router and the computer, the building materials used, interference from cordless phones and other devices etc. The data transfer rate is high at short range and decrease as the distance between equipment increases. If there is more than one user the data transfer speed can drop as the channel bandwidth is shared by all users

The data transfer speed is the connection speed between devices, actual internet data throughput depend on router connection speed to internet and congestion in public internet.

Wi-Fi operating range

Wi-Fi networks using the 2.4 GHz band (802.11b/g or 802.11n using 2.4GHz band) has a range of approximately 40m indoors and 100-500m (line of sight) outdoors. If the 5GHz band (802.11a or 802.11n using 5GHz band) is used the range is about one third of 2.4GHz band. The range depends on a number of factors like the type of building material used, obstruction encountered etc. The connection speed is higher when the connected devices are at close range. As the distance increases the connection speed decreases.

(A wide area Wi-Fi networks can have a wider coverage providing access within whole buildings, a series of multiple wireless access points (APs) are used to provide a wider coverage in those types of networks)

Wi-Fi logo

The name **Wi-Fi** (Wireless Fidelity) corresponds to the name of the certification given by the Wi-Fi Alliance, (formerly known as WECA -Wireless Ethernet Compatibility Alliance). Wi-Fi Alliance is the industry group which ensures compatibility between hardware devices that use the 802.11 standard. A Wi-Fi network, in reality, is a network that complies with the 802.11 standard. Hardware devices certified by the Wi-Fi Alliance are allowed to use the Wi-Fi logo

Wireless network configuration

There are several features in wireless networks and devices that need to be properly configured.

SSID

SSID is the abbreviation for Service Set Identifier. SSID is the name used to identify a wireless network. The SSID is a sequence of alphanumeric characters (letters or numbers) and is case sensitive. SSIDs have a maximum length of 32 characters. The SSID on computers and Wi-Fi devices can be set manually or automatically. Devices that need to communicate should be connected to the same SSID.

SSID is transmitted by a wireless network access point so that wireless devices searching for a network connection can discover it and connect to that network.

SSID broadcasting can be disabled which makes the network invisible for wireless devices scanning for available networks. When the SSID is not broadcast in a network the name of the network (SSID)

should already be known and manually entered to the wireless device to connect to that network.

Wi-Fi Encryption

Encryption is used to protect data transmissions in a network from being hacked or from unauthorised access. Data traffic is encrypted (scrambled) when leaving a wireless access point or a device and decrypted upon arrival at the receiving device. A common key (password) need to be entered at both ends of a link. The longer the key, the more difficult for someone to guess it or break it using various methods. In order to connect to a secure wireless network the security key or passphrase (passkey) is required.

The network is unsecure when Wi-Fi encryption is not used. Anybody with a Wi-Fi device within the range can connect to such a network since the security key is not required.

WEP (Wired Encryption Protocol)

This is the encryption method used by first generation wireless networking equipment.

WEP has been superseded by WPA and is rarely used now.

WPA (Wi-Fi Protected Access)

Weaknesses in WEP encryption method resulted in the development of WPA and WPA2, the level of security of WPA is much higher than WEP. WPA2 is more secure than WPA and has replaced WPA in most instances.

If devices do not support WPA it is better to use WEP than not using any encryption. Most wireless routers come with both WEP and WPA encryption capability.

MAC control access

MAC address of wireless access device can also be used to restrict access to a Wi-Fi network. Every Wi-Fi access device has its unique Media Access Control (MAC) number allocated by the manufacturer.

To increase wireless network security, it is possible to program a Wi-Fi network to accept only predefined MAC addresses and filter out all others. If a computer with an unknown MAC address tries to connect such a network, the access point will not allow it.

Access Point (AP)

Access points in Wi-Fi networks are similar to radio base stations in mobile cellular networks. The radio base station in a mobile cellular network maintains connections with mobile phones in the area using radio signals and connects through to the public phone network. Similarly, access points (APs) connect the wireless devices to the wired network. An AP both transmits and receives network data. It is usually connected to the wired data network through the use of an Ethernet cable. The AP, or the antenna connected to it, is generally mounted high on a wall or ceiling. It enables line-of-sight transmission to the Wi-Fi devices. APs coverage can range from 30m to 500m. ADSL/Cable modems with wireless routing functionality have a built in Access Point.

The number of users that can be attached to one AP at a time is normally limited by the processing capacity of the AP. It can range from a few users for a small low capacity AP up to 256 users for high end AP. When more users access the AP, the total data rate will be shared among all active users, resulting the data rate for each individual user being decreased.

CSMA/CA

CSMA/CA (Carrier Sense Multiple Access with Collision Avoidance) is the access control method in Wi-Fi networks. Wi-Fi devices usually cannot transmit and receive at the same time (not full duplex) and cannot detect collisions. CSMA/CA acts to prevent collisions before it happens. A Wi-Fi device wishing to transmit listens to the radio channels to see if the channel is clear (no other device is transmitting). If the channel is occupied the transmission is delayed for a random interval of time and an attempt is made again later to see if the channel is free. If the channel is free the Wi-Fi device will initiate transmission and other devices in the network refrain from transmitting in that time.

In a wired Ethernet Local Area Network (LAN) a method called CSMA/CD (Carrier Sense Multiple Access with Collision Detection) that attempts to detect collisions is used. Network interface devices are mostly full duplex (can transmit and receive at the same time) in wired Ethernet which makes it possible to detect collisions. A station will initiate transmission when the medium is free. When a collision is detected, transmission is halted for a random amount of time and another attempt made later on when the medium is free.

Wi-Fi Frequency bands

2.4GHz Band

2.40 -2.495 GHz radio frequency range is used. Each channel is 20MHz wide. A total of 14 overlapping channels numbered 1-14 are used within a 100MHz space. There is a 5 MHz separation between centre frequencies of these channels.

All these channels are not allowed in all countries. (Ch. 1-11 in USA, Ch1-13 in Europe and Ch. 1-14 in Japan are allowed). Only a maximum of 3 non overlapping (usually channels 1, 6, and 11) can be used at a time in one area.

5GHz band

24 non overlapping operating channels each 20MHz wide are available, total available bandwidth is 480MHz. The frequency allocation is from 5.170-5.835 GHz

(5.15-5.35 GHz and 5.725-5.825 GHz in USA). The frequency allocation range differs from country to country.

Wi-Fi transmit power

Wi-Fi device transmitter output power range from 40mW to 1 W depending on the frequency and country of use.

Wi-Fi Summary

Protocol	Freq. Range	Max. Data Rate	Typical Data Rate	Modulation Technique	Indoor Range (approx.)	Outdoor Range (approx.)
802.11a	5 GHz	54 Mbps	25 Mbps	OFDM	~ 35m	~ 120m
802.11b	2.4 GHz	11 Mbps	5 Mbps	DSSS	~ 40m	~ 140m
802.11g	2.4 GHz	54 Mbps	19 Mbps	OFDM	~ 40m	~ 140m
802.11n	2.4 GHz 5 GHz	300-600 Mbps	74 Mbps	DFSS or OFDM	~ 70m	~250m

Mobile Phones with Wi-Fi

Most new mobile phones are equipped with built in Wi-Fi feature enabling them to connect to Wi-Fi networks that has internet. In such instances calls can be made using VOIP features such as through Skype using the Wi-Fi connection.

Wi-Fi Direct

Wi-Fi Direct is a new specification that is being developed that would enable Wi-Fi devices to connect to one another without joining a traditional home, office, or hotspot network. With Wi-Fi Direct, devices will be able to make a one-to-one connection, or a group of several devices can connect simultaneously on peer-to-peer basis without going through a wireless router or wireless access point. Any Wi-Fi device, from mobile phones, cameras, printers, and notebook computers, to keyboards and headphones could be used with Wi-Fi Direct including those that are currently in use which were made before the Wi-Fi Direct specification. Devices without Wi-Fi Direct capability would require a software patch to enable them to connect using Wi-Fi Direct.

802.11ac

802.11ac also known as VHT (Very High Throughput) is the latest wireless networking standard that is being developed in the 802.11 series. 802.11ac will provide much faster data communication speeds than 802.11n and will operate in the 5GHz band. 802.11ac will be backward compatible with the earlier 802.11 series standards that used

5GHz band (802.11a and 802.11n). 802.11ac will use 80MHz channel bandwidth and has the option to use up to 160MHz channel bandwidth. An 802.11ac device will be capable of transferring data at a rate of 500Mbps when using an 80MHz channel bandwidth. This standard has the capability to use up to 8 spatial streams and supports higher order radio signal modulation up to 256QAM. MU-MIMO (Multi User – Multiple Input Multiple Output) antenna technology will be used here, one or more antennas will be used to transmit or receive independent data streams simultaneously. The higher throughput and data rates in 802.11ac are expected to be used in future for many applications. Upload and download of larger files from servers, campus and auditorium deployments, distribution of in home video content including HDTV are some of the services that are expected to use the high speed data transfer capabilities of 802.11ac.

15

ADSL

ADSL is the abbreviation for *Asymmetric Digital Subscriber Line.* (Asymmetric means the upload and download data speeds are different, not symmetric.)

ADSL service uses the wire line telephone connection for Broadband Internet access and data transfer. In an ADSL connection, Internet access is always available. The phone line can also be used simultaneously for voice telephone calls while browsing internet. Internet access speed using ADSL is many times faster than a dial up access. A flat monthly fee is usually paid for ADSL service and it enables 24 hour Internet access. Unlike in dial up access a separate phone usage charge is not applicable for Internet access.

Human voice in normal conversations can be carried within 0-3400Hz frequency range. The telephone networks were originally designed to use only this band in the frequency range 0 – 4000Hz for voice communication. Later it was discovered that the unutilized higher frequency portion can be used for data transfer. ADSL exploits this extra capacity to carry information by using the higher frequency bands for data transmission.

ADSL uses two frequency bands for upstream and downstream communications from telephone exchange to PC as shown in diagram.

A modulation scheme called DMT (Discrete Multi-Tone) modulation is used for the transmission of ADSL signals.

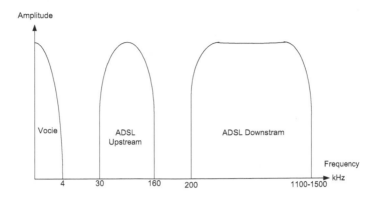

ADSL enabled telephone line frequency diagram

Uplink/Downlink speed

Uplink (or upstream/upload) means data sent from PC to Internet.

Downlink (or downstream/download) means data received from Internet to the PC

In ADSL the download speed is higher than upload speed.

When browsing Internet usually little information is sent upstream, when a hyperlink is clicked a few data packets are sent upstream to request a web page. But a lot of data is sent back downstream as the web pages which are downloaded to the PC containing a lot of data packets.

ADSL upload/download speed difference is therefore well suited for Internet browsing.

ADSL is capable of delivering 8Mbps download speed, uplink/downlink speeds for ADSL range from 64/256 kbps to 1/8Mbps, but the speed may be restricted by the ADSL service provider depending on subscription package.

(The maximum uplink/downlink speed in a V.90 dialup modem is 33/56 kbps)

Telephone line conditions

Since ADSL use the copper telephone line network for data transfer, the condition and type of line affects the ADSL service quality. Thicker gauge wires will facilitate higher data transfer and longer distances, while poor connections and contacts will hamper these. ADSL services can usually be provided for up to a distance of about 6km from the telephone exchange through telephone lines, the ADSL signal quality deteriorates with longer distances. Older and deteriorated telephone cables, corroded, poor cable joints severely reduce data download speeds in an ADSL connection.

Requirements for ADSL service

- A wire line telephone connection should be available to the premises

- The local telephone exchange should be equipped to provide ADSL facility.

- The distance from local telephone exchange to premises should usually be less than about 6km, if it is more, a line test will have to be done by the service provider to see if the service can be provided. As the ADSL signal quality drops with phone line length the distance should be within these limits.

- An ADSL Modem

If the above facilities are available ADSL services can be obtained after registration and paying the relevant fees.

ADSL Service subscription

ADSL service providers offer various ADSL service packages to customers. The charges for these packages depend on uplink/downlink speeds and the amount of data download limits. Basically there are three types of service packages offered by service providers in different countries.

1. Unlimited data downloads for a month at a given speed

2. Limited amount of data downloads for a month (1GB, 5GB, 50GB etc.) and additional charges for additional downloads beyond that limit within that month period

3. Limited amount of data downloads for a month (1GB, 5GB, 50 GB etc.) at a given ADSL speed and once this limit is exceeded the

download speed curtailed to dialup speed (usually 64kbps) or 256kbs till the end of the billing cycle.

ADSL Modems

ADSL modems are usually more expensive than dial up modems. There are several types of modems. ADSL USB modems will connect to the PC through USB port. ADSL Modem Routers will connect to PC through the Ethernet port with an Ethernet cable. Some ADSL modems have both USB and Ethernet connection facility.

If more than one computer need to be connected to a single ADSL line, ADSL modems with built in Router are available which usually will have 4 Ethernet ports.

ADSL modems with Wireless Router (Wi-Fi) facility are also available where the PC can access ADSL through the modem without having a wired connection to modem.

The ADSL modem should be properly configured with correct username and password provided by the ADSL service provider and connected to the phone line and computer

ADSL Splitters/filters

When using an ADSL enabled phone line, an ADSL filter/splitter should be installed in the phone line before every telephone or fax machine. This is to minimize the noise in the phone line and also to prevent disconnections. ADSL filters are also known as micro filters. Telephone wires were designed originally to carry voice signals which are usually under 4kHz frequency. ADSL uses frequencies very much higher than this speech band to carry fast data traffic. ADSL systems use typically frequencies between 25 kHz and around 1.1 MHz These higher frequency signals may result as noise in telephone calls and interfere with normal phone line operation. ADSL connection may tend to be disconnected when using telephone without an ADSL filter or splitter. The ADSL Splitter / filter isolates the telephone signal from ADSL signals.

ADSL Splitter is a small plastic box with an electronic circuit inside and has an input port which connects to the phone line. It has two output ports where one port is connected to the ADSL modem while the other port is connected to a phone.

ADSL filter has an input port to connect the phone line and an output port to connect a telephone.

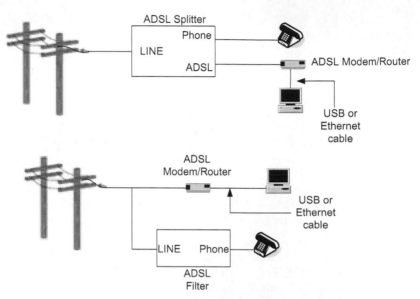

Two methods of ADSL connections using ADSL filter and ADSL splitter

ADSL 2

ADSL2 extends the capability of basic ADSL in data rates. The data rates can, in the best situation, be as high as 12 Mbps downstream and 3.5 Mbps upstream depending on line quality.

ADSL2+

ADSL services with higher speeds up to 24Mbps downlink and 3.5Mbps uplink are possible with ADSL2+ standards. The speed depends on various factors like distance from exchange, type of modem, phone line quality etc. ADSL2+ uses up to 2.2MHz phone line frequency band width which is double of typical ADSL frequency bandwidth of 1.1 MHz. To achieve higher ADSL2 or ADSL2+ speeds a relevant compatible modem is needed. Usually ADSL2 and ADSL2+ modems are backward compatible for use with standard ADSL.

Line filters for ADSL2+

ADSL 2+ need in line filters specially designed for ADSL2+, which supports a wider bandwidth. Although filters designed for ADSL will work with ADSL2+, expected ADSL2+ performance will not be achieved.

Connection Speed vs. Data Throughput

There is a distinct difference between ADSL connection speed and data throughput. ADSL connection speed is the connected bandwidth between the computer and the telecom exchange. That does not mean when browsing Internet all data will be downloaded at that speed. Download speed from Internet depend on various factors like the network and International Internet backbone congestion, bandwidth to the Internet from the server that is being accessed, number of people accessing the server at a given time etc. Data throughput is usually much less than the ADSL connected speed.

Naked ADSL

Naked ADSL is a term used to describe ADSL services provided by a telephone company through a telephone line but without giving a telephone connection. Few companies in some countries have started providing naked ADSL services and subscribers can get ADSL broadband services only without the need to pay line rental for a phone connection. Subscribers have the option of connecting an IP phone to the broadband router which may offer lower phone charges than a fixed phone.

XDSL

Unlike ADSL, DSL services provide same data transfer speed for both uplink and downlink. XDSL is the common term used to describe various DSL technologies like VDSL (Very High Speed Digital Subscriber Line), VDSL2, HDSL (High Data rate Digital Subscriber Line), HDSL2, SDSL (Symmetric Digital Subscriber Line). Telephone companies provide wire line based DSL services mostly for commercial and business applications.

16

Dial up Modems

A modem (modulator – demodulator) is a device that is used with a computer to send and receive data through a telephone line. A modem converts digital signals from computer to analogue signals to be sent over the phone line. It also converts the analogue signals received over the phone line to digital form to be fed in to the computer.

Computers can have built in modems or an external modem and can be connected to the serial or USB port of the computer. External modems usually require a separate power supply to operate.

A dial up modem with a computer can be used to access internet using a phone line. To access internet it is required to set the modem to dial an ISP (Internet Service Provider) and also to have an internet account with the ISP. When dial up modems are used to browse internet, normal phone usage charges are incurred for the duration of the connection.

Modern dial up modems can support data transfer speeds of up to 56kbps, although actual operating speeds are much lower and depend on telephone line conditions such as line noise, condition of line joints, length, type and length of lines and other factors.

Most dial up modems can also be used to send and receive faxes. Documents in a computer can be faxed using a dial up modem using appropriated fax software without the need of a separate fax machine.

Wider availability of ADSL, cable and wireless Broadband services have resulted in less number of people using Dial up modems for internet access.

17

ISDN

ISDN is the abbreviation for Integrated Services Digital Network. ISDN enables the transmission of digital voice and data over ordinary telephone lines. ISDN service is a digital telephone service for voice and data communications provided by wire line telephone service providers.

Before widespread availability of ADSL, ISDN was the method used to provide digital phone, data and internet services.

Ordinary telephone lines carry analogue signals. ISDN sends digital signals over normal phone lines. Data connection speeds up to 128 kbps both upstream and downstream can be provided using ISDN lines for a Basic rate connection, a type of connection usually used for non-commercial use. Two simultaneous connections over a single pair of line are available in any combination of data, voice, video or fax. Primary rate ISDN connections which are mostly used for commercial purposes can provide 2Mbps connections.

With a regular analogue telephone connection, the maximum data speed that can be attained using a dialup modem is 56 Kbps. With ISDN, a digital connection is established, with a higher speed. BRI (Basic Rate Interface) ISDN gives two 64 Kbps B (Bearer) channels.

The two channels can be combined together to achieve a speed of 128 Kbps.

ISDN is an end to end digital connection and can provide clearer voice and data. Analogue telephone sets convert the sound waves of our voice to analogue electrical waves (analogue transmission). In contrast ISDN equipment converts our voice into digital signals (voltages representing a string of 0s and 1s)

In BRI it is possible to make two simultaneous telephone calls over the same line. In ISDN it typically takes few seconds to make connections. Data can be sent more reliably and faster than with the analogue systems Noise, distortion, echoes and crosstalk are almost completely eliminated in an ISDN system.

ISDN can make and receive voice calls from normal telephone lines. To get the full benefit of high speed data transfer both ends of a connection should be ISDN lines with ISDN compatible equipment.

In most instances ISDN requires special equipment. When ISDN is being used for voice applications, a special digital ISDN telephone set is desirable to take full advantage of features.

There are two types of channels in an ISDN system.

B Channel – Bearer Channel

D channel – Data Channel

The B channels carry the payload data which may be voice, data or both voice and data.

The D channel is used for signalling and control. In some instances D channel is also used to provide low bit rate data connections.

There are two types of ISDN services, Basic rate (BRI - Basic Rate Interface) and Primary rate (PRI – Primary Rate Interface)

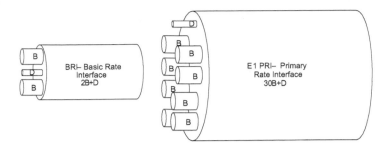

BRI

BRI can provide a data transfer speed of 144 kbps consisting of two 64kbps B (Bearer) channels and a 16kbps D or data channel. BRI is also known as 2B+D. A pair of standard twisted copper wire used for a telephone connection is used for a BRI connection.

PRI

PRI is mostly used for commercial applications. PRI can provide 30 B channels of 64kbps each and a D channel of 64 kbps. It can deliver a total of 2048 kbps PRI is also known as 30B+D or an E1 connection. In USA and Japan there is a slight different version where only 23 B channels and a D channel is delivered totalling a data rate of 1.544, which is also known as 23B+D or a T1 connection. A PRI connection uses four wires, one pair for data transfer in each direction.

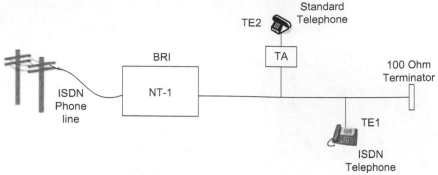

ISDN Connection Diagram

TE1 – ISDN devices

TE2 – Non ISDN devices

TA- Terminal Adaptor – Used to interconnect non ISDN devices to ISDN line

NT1 - Network Interface for BRI services

(For PRI services Network Interface is known as NT2)

Digital equipment like digital telephones, digital (G4) fax machines can be connected to ISDN network. A terminal adaptor (TA) is used to convert the digital signal to an analogue signal when connecting analogue equipment such as ordinary telephone and fax machines to an ISDN line.

ISDN charges are usually similar to voice call charges and are based on duration of call. Charges are incurred for each line for a BRI connection if both B channels are used.

ISDN has been mostly replaced by affordable ADSL services in most parts of the world.

18

Leased Lines

A Leased line is a dedicated telecommunication line connecting two locations or a reserved wire line circuit between two locations. Leased lines do not have phone numbers associated with them and are not connected to the PSTN (Public Switched Telephone Network). Leased lines are also known as private lines or tie lines. Leased lines facilitate symmetric (same speed for uplink and downlink) data transfer and provide a permanent connection between two different locations. Leased lines can also be used for internet access if they are connected to an ISP (Internet Service Provider). Leased lines are generally used for data connectivity, to connect PABXs, or to connect computer networks in different locations. Earlier analogue connections were provided for leased lines but most leased lines provided now are digital connections.

Leased lines are provided by Telecommunication Network operators. Twisted pair copper lines, coaxial cable, fibre or a combination of these can be used to provide leased lines. Wireless network operators can provide dedicated connections with connection speeds equivalent to leased lines using wireless connections. An installation fee and a flat monthly fee are usually charged for leased line connections depending on the speed or bandwidth of the connection and distance.

Digital leased lines are provided at speeds of 64kbps and multiples of it. A 2.048Mbps connection is known as an E1 connection. E1 has 32 channels of 64kbps (two of the channels are used exclusively for synchronisation and signalling purposes). In USA and Japan a different configuration named T1 is used which gives a 1.544Mbps connection consisting 23 channels.

Leased lines are also used by businesses that need high speed reliable Internet access, such connections offer dedicated, uncontended bandwidth ensuring that the maximum line speed is always available when required. Guaranteed bandwidth can also be obtained direct to the internet backbone using leased line internet connections.

Dedicated links equivalent to leased lines are also provided by wireless telecom service providers using wireless links.

19

Fibre Optic Communication

Optical Fibres are long thin strands of glass which are about the diameter of human hair. Optical fibres are used to carry information in the form of light signals over long distances. A fibre optic cable consists of hundreds of thin strands of optical fibre which are bundled together. Unlike in electrical cables which carry information in the form of electrical signals, fibre optic cables can carry massive amount of information in the form of light signals. Another advantage of fibre optic communication is they are immune to electromagnetic interference since they carry information in the form of light waves.

Telephone, data and internet networks today use high bandwidth fibre optic cables to carry huge amount of information. Telecommunication backbone networks which interconnect telephone exchanges and data centres use fibre optic links for interconnections.

Fibre Optic submarine cables

Fibre optic submarine cables laid on seabed under the oceans are used to interconnect telecommunication networks in different countries and carry massive amount of data. Specially designed cable laying ships are used to lay these cables on sea bed. The cables are connected to telecommunication networks through gateway stations at landing points on coast.

93

Dark Fibre

Dark Fibre or Unlit Fibre is the term used to denote fibre optic cables that have been laid but still not being used. When fibre optic cables are laid, additional cables are often laid for future use; these are not connected to fibre optic transmitters, receivers or networks and are known as dark fibre.

Wavelength Division Multiplexing (WDM)

WDM is a multiplexing technology where light waves of different wavelengths are sent in the fibre effectively multiplying the data carrying capacity by many times. It is similar to sending light waves of different colours. This allows a fibre to increase its capacity and looks like there are multiple fibres to carry different streams which in effect are done by a single fibre. DWDM (Dense Wavelength Division Multiplexing) is a technology which carries even more number of data streams at different wavelengths than WDM and is being increasingly used to improve the data carrying capacity of fibre optic cables.

Fibre to the Home (FTTH)

Many countries have already started laying out FTTH networks where each individual home gets a fibre optic connection to have a high speed broadband connection. In some instances FTTN (Fibre to the Node) networks terminate fibre cables in neighbourhood cabinets and homes are connected from the cabinet through DSL or Coaxial cables to give high speed broadband access. FTTH/FTTN networks will enable connection speeds of up to 100Mbps or more for homes and in addition to broadband access will facilitate many applications like IPTV and Video conferencing which require reliable high bandwidth access.

20

VoIP

VoIP is the abbreviation for Voice over Internet Protocol. It is a method of using the internet to talk to a remote party or to make voice phone calls to a remote party. VoIP converts analogue voice signals in to digital data packets and send them over the internet where at the receiving end the digital data packets are converted back in to voice signals. Most of the VoIP applications can be used free of charge and VoIP is a popular method to speak to people across countries at a low cost or completely free of charge when internet access is available.

To use VoIP, a connection to the internet is required. If using a computer, a suitable sound card, a microphone, a speaker/headset and appropriate VoIP software are also needed. USB VoIP phone devices that can be plugged in to a USB port of a computer can also be used instead of a mic/headset with a computer to make VoIP calls.

Packet Switching

Traditional telephone calls use circuit switching technology where a dedicated circuit is set up for the whole duration of the call between the two parties in conversation. VoIP uses packet switching technology where voice signals are converted in to data packets and sent over the internet. Different data packets can take different paths through internet to reach the intended destination. At the destination all packets are reassembled in correct order and digital data are converted in to voice signals. Advantage of packet switching is that the amount of bandwidth occupied during a call is a fraction of what is required for a circuit switched call. Disadvantage of VoIP is the call quality may be lower than direct dialled circuit switched calls due to packet losses and delays in data packet transfer.

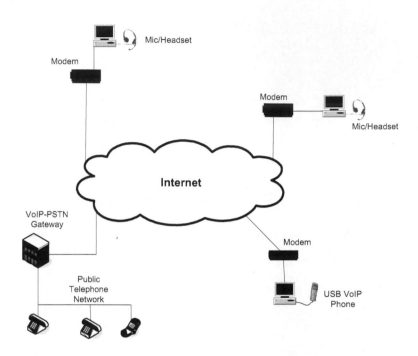

Most of the VoIP calls need two computers at each end connected to Internet. Other devices can also be used instead of computers to make VoIP calls. IP devices and telephones that directly connect to broadband routers are also available for VoIP applications and a computer is not needed in such instances. VoIP calls can also be made from mobile phones which have a wireless data connection by installing required software or apps in the phone.

Skype, Yahoo, MSN are popular VoIP applications. Opening an account using any of these applications is free of charge and will enable one to make VoIP calls. Calls can be made to ordinary telephone lines at a lower cost than international phone charges with these VoIP applications Most of these applications allow a video conversation when a web camera is installed.

VoIP - PSTN Gateways

VoIP calls can be made to normal phone lines and mobile phones when the VoIP network is connected to the PSTN (Public Switched Telephone Network) through a VoIP – PSTN gateway. Usually calls made to ordinary phones and mobile phones are not free and a paid account with the VoIP service provider is required to make such calls.

SIP telephone services also use VoIP and do not require a computer to operate, if a hardware SIP device is available, only an internet connection is required. SIP services are described in a separate chapter.

Cheap International Calls

Some of the mobile and fixed phone companies offer international calls using VoIP at a cheaper rate than international direct dialled (IDD) calls. International phone calls made at a cheaper rate using phone cards are most often sent using VoIP. In these instances calls connect to the phone exchanges using circuit switching the usual way and the international part of the call to the foreign country is sent using VoIP. Due to packet loss and packet transfer delays call quality may be lower in VoIP calls than IDD calls.

21

Metro Ethernet

Ethernet technology standard is the most widely used standard to interconnect computers in a local area computer network. Almost all LANs (Local Area Networks) use Ethernet protocol. A Metro Ethernet network is a network spanning over a metropolitan area and provide high speed (bandwidth) connectivity to commercial buildings. It is used to interconnect corporate computer networks using the Ethernet standard.

Metro Ethernet can be used to connect networks on a Point to Point or Point to multi Point basis and for High Speed Internet access.

Metro Ethernet networks are deployed by a Telecom service provider. To obtain connectivity, subscription charges have to be paid for the service provider. Metro Ethernet is generally used for commercial and business purposes. Up to 10 Gbps data speeds could be provided depending on the network and subscription service.

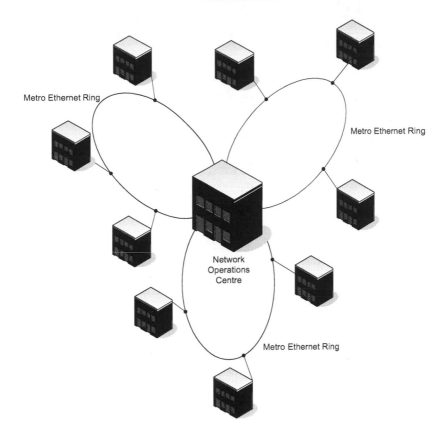

Four different versions of Metro Ethernet services are commonly used.

- MPLS (Multi Protocol Label Switching) based Metro Ethernet

- Pure Ethernet based Metro Ethernet

- SDH (Synchronous Digital Hierarchy) based Metro Ethernet

- DWDM (Dense Wavelength Division Multiplexing) based Metro Ethernet

Some of the services that can be offered over Metro Ethernet include the following.

- High Bandwidth Internet access

- Virtual Private Local Area Network Services - This allows two or more Local Area Networks situated at different sites to be connected over Metro Ethernet connection to form a Virtual LAN.

- Virtual Leased Line – A point to point connection for data transport

Metro Ethernet connections can be provided using fibre or copper cables or a combination of both fibre and copper.

22

SIP (Session Initiation Protocol)

SIP is a signalling protocol designed to establish sessions in an IP (Internet Protocol) network. A session can be a two way telephone conversation over internet or a collaborative multimedia conference session. SIP protocol is used to establish, terminate and modify communications sessions in an IP network.

To use SIP telephone services, a broadband internet connection and connection to a VoIP SIP service provider or a SIP server a is required. Access to these enables one to receive and make SIP telephone calls using the internet.

SIP Phones are also known as VoIP Phones or soft phones. These are telephones that allow phone calls to be made using VoIP (Voice over Internet Protocol) technology. In SIP phones, voice is converted to digital signals and compiled in to data packets that are sent over the internet.

SIP uses Peer to Peer (P2P) feature for connections. A SIP server facilitates the establishment of a connection. Once the connection is established, communication takes place directly between two connected parties using P2P without going through the SIP server.

SIP is gaining more popularity and gradually replacing H.323 protocol which has been used to send voice over internet.

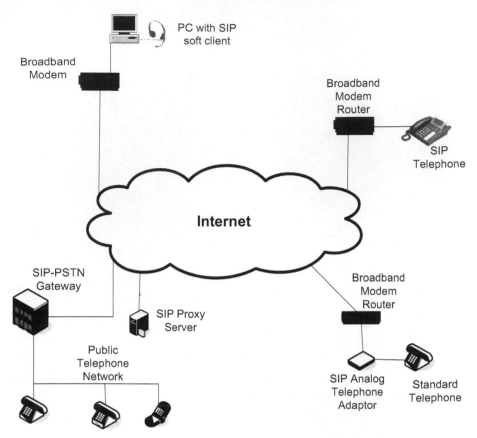

Different methods of SIP connections

There are two types of SIP Phones.

Hardware SIP Phone

The first type is the hardware SIP phone, which is connected to a broadband router. Alternatively a SIP ATA (Analogue Telephone Adaptor) can also be used. A SIP adaptor is connected to broadband router and it also has a phone socket to connect an ordinary telephone. Access to a broadband router with a broadband connection is required when using a hardware SIP phone to make calls and a computer is not needed.

Software SIP Phone

SIP Phones can also be software-based. A software based SIP phone is a software program which makes use of computer's soundcard with microphone and speakers, or an attached headset to make or receive calls. SIP Phone client software is required to be installed in the PC.

To use SIP service the following are needed.

- A broadband connection

- A SIP phone (hardware or software)

- A user account with a SIP service provider.

Advantage of a SIP Phone

Most SIP phones can also make calls to normal landline phones or mobile phones through a gateway connected to a SIP Server. Calls to ordinary phone lines through a gateway may incur a charge.

A SIP user who has obtained a SIP user account from a SIP service provider has the ability to use that account from anywhere a broadband connection is available. A telephone number need no longer be restricted to one particular geographic location. It can be used in a physical SIP phone or in a PC as a soft phone.

A SIP phone can be used just as a normal phone. A computer is only required for a software SIP phone.

For example a SIP user who obtains an Australian phone number can use that account from anywhere in world. If he uses his account in USA, he can contact other Australian phone numbers at local call rate. In many instances calls between two users of same SIP service provider (SIP to SIP calls) are free of charge irrespective of where they are located.

23

FM Radio

FM is the abbreviation for Frequency Modulation. FM is the Modulation scheme used to broadcast Hi-Fidelity audio which is popularly known as FM Radio.

FM Radio Broadcasting uses the VHF (Very High Frequency) radio spectrum of 87.5 - 108.0 MHz. In Japan FM Radio band uses 76 - 90 MHz frequency band.

In FM, the frequency of the carrier wave is varied according to the frequency variation of the input signal. In AM (Amplitude Modulation) the amplitude of the carrier wave is varied according to the frequency variation of input signal. (Both Short Wave – SW and Medium Wave – MW Radio Broadcasts use Amplitude Modulation)

FM Radio is used to broadcast Hi-Fi music and speech over the air since it is more immune to noise and interference from other stations. It also enables to provide richer quality audio broadcasting. The audible musical range of a human ear is about 20Hz - 20kHz. FM radio channel is wide enough to accommodate this range; the reason FM radio sounds better quality than AM radio. Compared to AM Radio, FM Radio uses more bandwidth per channel. AM Radio uses 5-10kHz bandwidth.

FM Radio channel bandwidth

The maximum deviation of an FM Radio channel is 75 kHz to either side of the centre frequency. Deviation is the degree of variation of the carrier signal from the centre frequency. Each channel has a 25 kHz upper and lower guard band. The bandwidth effectively is 150kHz and with two upper and lower guard bands of 25kHz each channel occupies 200kHz bandwidth. FM Radio stations are assigned frequencies with 200kHz separation. Theoretically it is possible to have 100 different Radio channels within the VHF FM radio broadcasting band. Most new advanced FM radio receivers indicate the tuned frequency in a digital display.

The coverage distance from an FM radio broadcast station is less than that of an AM station due to the high frequency band used and the wider channel bandwidth. The range is dependent on transmitter station power, antenna location, height and antenna gain. The range for stereo transmission is less than mono transmission. But superior audio quality has made FM Radio the preferred audio broadcasting method.

FM Stereo

FM Radio enables the broadcasting of Hi Fidelity Stereo Music. Radio receivers with FM Stereo reception ability can receive FM Stereo transmissions with left and right side speakers giving independent audio outputs.

FM Stereo transmissions include a 19 kHz pilot tone signal embedded in the signal. FM stereo receivers upon detecting the 19 kHz pilot tone will switch on the FM stereo decoder unit. Most FM Stereo receivers will also have an FM Stereo indicator lamp to indicate the reception of an FM stereo transmission. The FM stereo decoder unit will decode the received signals and output left and right channels separately.

Mono FM radio receivers can also receive stereo broadcasts and the combined left and right channels are given out from a single speaker. To properly receive stereo broadcasts the signal strength of received signal should be of a higher level whereas a weaker signal can give mono reception. Most stereo receivers have a stereo/mono switch, when signals are weak for stereo reception the switch can be set to mono position and the audio is output in mono format with left and right channels combined together. An external antenna or an extended telescopic antenna can greatly improve the reception of FM transmissions. In Analogue Television broadcasting the audio portion of the television signal is broadcast using FM modulation scheme.

Squelch

Squelch circuits in FM receivers are used to switch off the audio when a sufficiently strong signal is not received. This prevents unwanted noise being heard when the radio is switched on and not tuned to a station or when tuning between stations.

RDS

RDS (Radio Data System) is a feature used in FM Radio to send information related to the radio programme, traffic and travel news and to enable car radios to auto tune to a specific channel when moving from one area to another on a long journey.

A 57 kHz subcarrier embedded in the radio channel sends data in digital forms in RDS. Not all stations provide RDS. When a radio channel uses RDS, a receiver with RDS facility can use the information broadcasted related to radio channel.

RDS feature can be used to send information related to the radio channel to be displayed on the screen of the radio receiver. The information displayed can include radio station name, program details, song titles, current time etc.

RDS feature can also be used to broadcast travel and traffic related news. When a radio equipped with RDS is set to receive travel news, upon the reception of a travel or traffic announcement from a radio broadcast station, the receiver will temporarily pause the normal radio programme or a CD that is being played to allow the announcement to be heard.

RDS feature also has the ability to auto tune to a radio station. This feature is specially useful on long journeys where a car moves from the service area of a transmitter to another transmitter of a radio network channel. Without RDS the car radio will have to be manually tuned to the radio channel frequency of area. Some Radio networks broadcast a PI (Programme Identification) code which enables radios equipped with RDS to automatically tune to the same radio channel when moving from one area of transmitter coverage to another.

24

Digital Audio (Radio) Broadcasting

Digital Radio Broadcasting popularly known as Digital Audio Broadcasting (DAB) is the broadcasting and reception of Digital Radio services. To receive digital radio services, a digital radio receiver is needed and a digital broadcasting service should be available in the reception area.

In Digital Radio broadcasting, audio is converted in to digital format before being broadcasted. A Digital Radio receiver that receives the digital radio signals converts them to audio signals for listening.

In Digital Radio Broadcasting, Digital Signal Processing (DSP) methods are used to compress and manipulate the broadcasted signal. Error Correction techniques are used to ensure that the errors in received signal are minimised.

Compared to analogue radio broadcasting techniques such as AM and FM, Digital Broadcasting offer many additional features and advantages. More digital radio channels can be accommodated in the radio frequency spectrum in the space occupied by an analogue channel. Several channels can be multiplexed in to a single frequency and broadcast. Digital radio offers better quality audio and is less susceptible to noise, fading and interference than analogue broadcasts.

Another advantage of digital radio is the transmitters can use the same frequency in different geographical areas in a network without interfering. This eliminates the need for a listener travelling in a vehicle to retune a radio channel when moving to a different area.

DAB features

Digital radio receivers have small screens to display text and simple graphical information. Electronic Programme Guide (EPG) is a feature of digital broadcasting where listeners can see information about programmes being broadcast. It can also include information about a specific programme, news or weather alerts etc. on the screen. Tuning to a digital radio channel is easier as listeners can select a channel or station from the names displayed in the screen of the receiver.

Digital Radio broadcasting services have been started in many countries. Initially both analogue and digital channels are broadcast together (simulcast). In some instances digital only channels are also being broadcast. In future more digital channels are expected to be available. More digital radios are also expected to be available in the market and the prices of digital radios are expected to come down with the availability of more digital channels.

DAB+

Digital radio started with DAB standard. DAB+ is the latest digital audio broadcasting standard which is superior to DAB with higher quality signals. Earlier DAB standard radios are not compatible with newer DAB+ standard; some of the DAB receivers may be upgradable to DAB+ by a firmware upgrade.

DAB uses MPEG Audio Layer II which is also known as MP2 codec for audio compression and decompression. DAB+ uses MPEG-4 audio codec also known as MP4 or HE AAC+ (High Efficiency Advanced Audio Coding version 2) which allows better audio quality to be broadcast using lower bit rates.

DAB modulation and radio frequency

The radio frequency modulation technique used in DAB is COFDM (Coded Orthogonal Frequency Division Multiplexing).

174 - 240 MHz (VHF band III) and 1452 - 1479 MHz (band L), are the frequency bands assigned for DAB+. Different countries have allocated different portions of these frequency bands for Digital Radio broadcasting. Some countries have reserved the upper portion of L band for satellite delivery of digital radio.

DAB+ receivers

HD Radio

HD Radio or High Definition Radio is the standard used in USA for Digital Radio broadcasting.

Digital versions of FM and AM channels are broadcast in digital HD radio format in the USA. Some AM and FM radio stations simulcast both analogue and digital signals known as a hybrid digital-analogue signal within the same channel. Most HD radio receivers have the ability to switch to analogue format signal if the digital signal is not received or lost.

25

Satellite Radio

Satellite Radio service involves the transmission of radio broadcast services from a broadcast communication satellite in space and the reception of its signals using a satellite radio receiver.

Satellite Radio services can cover a larger area than terrestrial (over the air) radio and reception of a satellite radio channel is possible over several countries and continents.

There are many satellite radio channels available, most of which are subscription based channels. A number of free channels are also available. Some of the satellite radio channels are also simulcast as streaming audio on Internet.

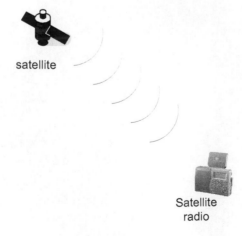

satellite

Satellite
radio

Many models of Satellite Radios are available for home, vehicle and portable use. Satellite Radios have proprietary chipsets and are specifically designed for different satellite radio services. They must be purchased specifically for the intended satellite radio service that needs to be received.

Sirius XM

Sirius and XM were two separate satellite radio services which were merged in July 2008 and the company is now known and Sirius XM but the services are operated separately. Sirius uses 3 satellites. XM uses four satellites for their services. 2.3GHz band is used for satellite signal transmission.

Sirius and XM also have a repeater network that re-transmits the signals received from satellites to Sirius and XM radios. The repeater network is used to ensure uninterrupted reception in urban areas where tall building and other obstruction may obstruct the direct satellite signals.

Sirius and XM cover USA and Canada. Sirius has 130 channels and XM has over 170 channels.

Audiovox SiriusXM Satellite Radio Receiver

1Worldspace (No longer operational)

1Worldspace (earlier known as Worldspace) was the only satellite radio service outside USA, Japan and South Korea. 1Wordspace was operational till end of 2009 and its services were terminated on account of 1Worldspace being bankrupt. Efforts were being made to relaunch the services.

1Worldspace used two geostationary satellites Afristar1 and Asiastar to broadcast its services. Afristar satellite provided coverage for Mediterranean, Middle Eastern and European regions. Asiastar coved southern Russia, South and South East Asia and China. Three separate beams of 50 channels each were transmitted for different regions. 1Worldspace used L-Band frequency (1452-1492 MHz range) for broadcasting. To receive 1Worldspace broadcast, a 1Worldspace satellite receiver with special chipset was needed. There were many models available including lightweight portable ones weighing 1.5kgs upwards.

Hitachi 1Worldspace satellite radio

Over 60 digital audio channels were available and the number of receivers in use worldwide exceeded 170,000, most of which were in India. Some of the channels were provided by third parties and include CNN and BBC.

(Satellite radio described here is different from another format of satellite radio called DTR – Digital Television Radio, where an audio only satellite radio channel can be picked up by a satellite TV receiver. This format is part of a satellite TV setup where there are a number of channels that broadcast audio only without the pictures and involve the use of a dish antenna and a satellite TV receiver.)

26

Television

Television transmission where signals are transmitted from a television transmitting station and received by a Television receiver using an antenna is known as Terrestrial (over the air) Television. (There are other methods of receiving television also using satellites, cables etc.). This chapter explains the commonly used analogue terrestrial television broadcasting system.

VHF and UHF bands

Television signals are broadcast using VHF and UHF radio frequency bands. VHF signals can cover a vast area from a television transmitting station than UHF signals. There are a greater number of channels in UHF bands than in VHF bands. UHF signals give better reception in urban areas with many buildings.

Television Transmission Bands and Channels

Band	Channel Numbers	Frequency
VHF Low (Band 1)	2- 4	47 – 68 MHz
VHF High (Band III)	5-12	174 -230 MHz
UHF	21 – 69	470 – 862MHz

(There can be slight variations of above channels and frequencies in different countries)

Video standards

There are three main video broadcast standards, PAL, SECAM and NTSC. These standards are incompatible with each other and depending on the country of use television receivers and video recorders should support the relevant standard. There are an increasing number of modern television receivers and video recorders that support multiple standards. Different systems have different number of lines per picture frame and number of frames per second.

NTSC (National Television System Committee)

This system is used mainly in American continent, Japan and some other countries.

Lines per frame – 525,   Frames per second -30

PAL (Phase Alternating Line)

This system is used in most European countries (except France) and many other countries. PAL system has various subdivisions, B,G,H,I,D,N etc.

Lines per frame - 625,   Frames per second – 25 (Except PAL M)

SECAM (Sequential Couleur Avec Memoire)

This system is used in France and a few countries.

Lines per frame – 625, Frames per second – 25

Television Signal Modulation

A television channel signal consist of two signals, video and audio signals. The picture information in a television signal is transmitted using VSB (Vestigial Sideband Modulation) which is a form of AM (Amplitude Modulation) while audio information is transmitted using FM (Frequency Modulation). The colour information is carried using a QAM (Quadrature Amplitude Modulation) subcarrier signal. SECAM uses FM modulation for the video signals. In PAL B each channel is allocated 7MHz of frequency bandwidth, while in NTSC 6MHz is allocated. Within this band video, audio signals and colour subcarrier signals are transmitted at different frequencies. Video signal carries the picture information and audio signal carries the sound. Audio signal is always 4.5MHz above the video signal frequency. Colour subcarrier carries the colour information of video signal, monochrome (Black & White) Televisions ignore this signal. Colour Television receivers decode this signal to extract the colours of the picture.

Digital Television Switchover

Many countries have already started Digital Television transmission and are phasing out analogue Television transmission. Digital Television has many advantages over analogue Television. Digital Television is explained in a separate chapter.

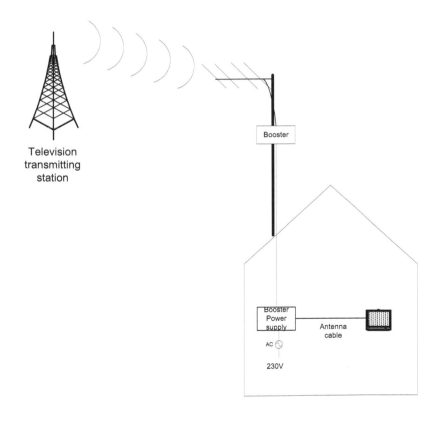

Television
transmitting
station

Booster

Booster
Power
supply

Antenna
cable

AC

230V

Television Reception

For proper reception of terrestrial television signals a well configured antenna system should be in place. The reception quality depends on various factors like the location, distance to the television transmission station and blockages to signal path, antenna type, height and orientation, antenna cable type and length etc.

Antenna Cable

Most television receivers use a coaxial cable to connect an antenna to the television. The impedance of cable should be 75 ohms. A good quality coaxial cable is essential to minimise signal loss. Usually thicker cables and cables with more strands are of better quality and results in low signal losses giving good picture quality. The length of the cable is also an important factor, the longer the cable the higher the losses resulting in low quality pictures. The length is a critical factor in UHF band signals since higher UHF frequency losses are more than in VHF bands. The location of the antenna should be such that a minimum cable length can be used to connect the antenna to the television. Cable connectors should also be of good quality and properly connected.

TV Antenna

A good quality antenna pointed towards the television transmitting station ensures good reception quality.

UHF antennas are different from VHF antennas due to their element lengths. There are some antennas with both UHF and VHF elements. If separate antennas are used for UHF and VHF, a combiner or a booster with combiner should be used to connect the two cables from UHF and VHF antennas in to a single cable.

Antenna Gain

Antenna gain is the ability to grasp television signals, the higher the gain the higher the ability to grasp weaker signals. Higher gain antennas have more antenna elements. Higher gain antennas also have a narrower beam width which means they can grasp signals coming within a narrow angle of the pointed direction and therefore have to be precisely pointed towards the transmitting station. If multiple channels are to be received from different television transmitting stations located

116

at different angles, having a higher gain antenna can be a disadvantage as it would be difficult to get signals from stations other than the pointed direction.

TV Signal Booster

Television signal booster is a device used to amplify weak television signals. The booster comes in two parts, the actual booster unit and the power supply unit. The booster unit should be fixed to the antenna mast and as close as possible to the antenna. This is to prevent further losses in the antenna cable to the weak signals received before they can be amplified. The booster power supply unit is plugged to the mains power and connected to the antenna cable before it is connected to the television. The antenna cable takes the DC power from the power supply unit to the booster unit fixed on the antenna mast. (DC power in the coaxial cable does not affect television signals as it does not mix with television signals in the cable).

The gain of a booster is the amount of amplification it can make to the received signal. A 3dB gain means amplification by a factor of 2, 6dB is amplification by a factor of 4 and 9dB is amplification by a factor of 8 and so on. Amplifying a signal too much can also result in bad quality pictures, therefore in an area of moderate strength television signals it is not wise to use a high gain booster. Some boosters have the facility to control the gain by a variable gain controller.

TV Tuners for computers

It is also possible to watch Television on a computer screen by attaching a TV tuner to the computer. TV tuners for computers are available as USB type units or PCI tuner cards. Some computers have built in TV tuners. Television programmes can also be recorded on the computer hard disk when TV tuners are used with computers. Most of the TV tuners available for computer are of digital type, to watch analogue television transmissions the TV tuner should support analogue Television.

Television additional features

NICAM

NICAM stands for Near Instantaneous Companded Audio Multiplex.

NICAM system is used to transmit stereo sound in analogue Television Broadcasting. A TV receiver equipped with Nicam decoder can receive stereo audio when a Nicam signal is transmitted. TVs without Nicam feature will receive a mono signal.

This can also be used to transmit two separate mono channels or audio in two languages where the user can select the desired language.

Teletext

Teletext is a text based information broadcast system using Television. Teletext service is popular in UK and European countries. It is used to broadcast national, international news, sports weather, stock market information, TV schedules etc. Teletext information consist of a number of pages of textual information

Teletext information is transmitted using the spare capacity in the television composite video signal. Teletext signals are embedded in the vertical blanking interval (VBI) of the television signal.

If a Television broadcaster offers a teletext service, a TV receiver equipped with teletext decoder can display the information one page at a time. TV Remote control is used to select the required pages for display.

A Teletext information page

118

Subtitles

A special feature of teletext is the ability to broadcast subtitles. Textual version of dialog or other information can be displayed at the bottom of the screen as subtitles. They are specially useful if the dialog is in a different language. It also helps those with impaired hearing to follow up the dialog.

Unlike in normal television picture that carry subtitles, when sent though teletext the user can switch subtitles on or off. Some programs might offer subtitles in different languages where the user can select the desired language.

27

Digital Television

Digital Television transmission involves the use of digitally encoded and compressed signals to broadcast and receive Television signals. There are many advantages in Digital Television compared to traditional Analogue Television. In Digital Television channels, pictures are of superior image quality and sound are of higher audio quality than in Analogue Television channels. Digital channels take less radio frequency bandwidth than analogue channels and can offer more number of channels for the same bandwidth occupied by traditional analogue television. Multi channelling offering more than one program on same channel and electronic program guides are features of digital television. Widescreen pictures, surround sound, HD (High Definition) pictures, ghost free reception, interactive services are some of the additional advantages of Digital Television.

Digital Television Standards

There are several Digital Television standards that are in use today.

DVB-T (Digital Video Broadcasting – Terrestrial) is the most widely used standard. It is used in Europe, Australia and many other countries. North America uses ATSC (Advanced Television Standards Committee) standard. ISDB-T is the standard used in Japan, China uses DMT-T/H standard.

In many countries around the world digital television transmissions have already been started. Initially most countries will have both analogue and digital versions of the same channel. Gradually the analogue channels will be phased out in future.

Some countries have already announced the dates when all analogue channels will be completely phased out and digital only channels will be on air.

In analogue television, transmitters in adjacent areas could not use the same channel frequency due to interference. OFDM modulation scheme used in digital television allows a single frequency or channel to be used without interference in adjacent areas. This enables all transmitters in different areas of a country to operate in a single channel or frequency.

To receive digital television, a digital television receiver is needed. A digital television receiver has a built in digital tuner. Most digital television receivers can receive analogue television channels as well. It is also possible to watch Digital TV on a computer screen with a Digital TV tuner attached to the computer. Computer Digital TV tuners are available in USB plug in format and as PCI cards for desktop computers.

Analogue television receivers cannot receive digital television channels since they do not have built in digital tuners. But an optional digital tuner set top box can be used together with an analogue television to watch digital television channels.

Digital Tuner Set top Box

Digital tuner set top box is a device that converts digital television signals to analogue format signals or video signals. The device receives signals through the antenna and converts them to analogue television signal format which can be viewed with an analogue TV. When the digital tuner set top box is used, changing of channels has to be done in the set top box and not in the television set. (In most instances Digital Set top box will have a remote control unit).

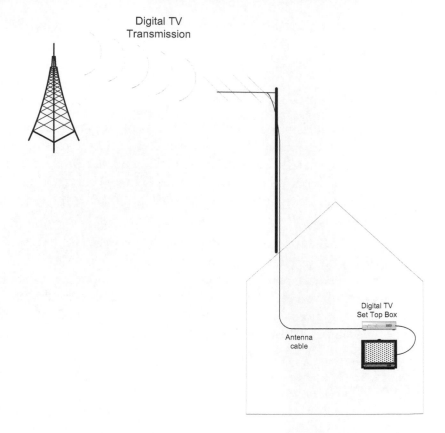

Digital television usually occupies a 6-8MHz wide band in the UHF or VHF frequency spectrum. Digital data transfer rates for digital channels can vary depending on the compression and definition (HD or SD) used. A HD Television channel will have a peak digital data rate (using MPEG compression) of about 28Mbits per second.

HD/SD

Digital television programmes can be either of High Definition (HD) or Standard Definition (SD) format. Definition refers to the level of resolution on screen. Resolution of a TV is made up of a number of lines and pixels. Higher the number of lines and pixels makes the image clearer and sharper. To receive HD programs the digital TV tuner should have HD decoding feature and the television monitor should be of HD type to display HD pictures. HD can offer cinema quality pictures and Dolby Digital Surround sound. HD pictures are of much better quality than SD since the number of lines per frame is about twice that of SD. HD Digital television pictures can also be of wide screen format with width to height or aspect ratio of 16:9 which is about one third wider than the 4:3 aspect ratio used in analogue television. Not all programs broadcast on digital channels are of HD format, they may be of SD format. HD format programmes should have been recorded in HD format using HD cameras. Usually at the beginning of a HD program it is indicated on screen that the program is available for viewing in HD format. There are some televisions that are marked HD Ready and unless specified these do not usually have a built in HD digital tuner. An optional HD digital tuner set top box may need to be connected to watch HD channels in these types of television sets.

A Full HD TV has a resolution of 1920X1080 which is made up of 1080 progressively scanned lines denoted as 1080p. Usually an HD ready TV means it does not have a built in HD tuner but the television can display HD pictures from an external source such as a DVD, Blue Ray player or HD Digital Television or Satellite television set top box.

MPEG

MPEG stands for Moving/Motion Picture Experts Group which develops standards for digital audio and video formats. By compressing, it is possible to reduce the bandwidth used for transmission or the space used for storing digitally encoded moving pictures. There are several versions of MPEG. MPEG-2 is commonly used to broadcast digital television. MPEG-4 is also increasingly being used which takes less bandwidth to deliver high quality pictures. Most HD channels are transmitted using MPEG-4 coding.

DVB-T transmissions can be coded with either MPEG-2 or MPEG-4.

To receive a digital television transmission encoded with MPEG-4, the digital television receiver or digital television set top box should have the MPEG-4 decoding facility.

Antennas for Digital Television

Television antennas used for analogue television can also be used to receive digital television signals. Technically there is no difference in antennas used for analogue or digital channels unless there is a change in the channel or frequency. If a different frequency or channel is used for digital television an antenna optimised for that frequency channel may be needed.

DVB-T2

DVB-T2 is the new technology standard evolving from DVB-T standard. It is expected to provide additional features and better performance than DVB-T. DVB-T2 is not expected to replace DVB-T services in short term but both systems are expected to co-exist with each other.

3D TV

Television receivers with 3D (3 Dimensional) feature let viewers experience movies, games and other video content with stereoscopic (three dimensional) effects. Ordinary Televisions display only 2D images which are two dimensional, width and height only. 3D displays add the illusion of a third dimension of depth.

3D capable displays together with specially designed eye glasses need to be used to get the effect of 3D. If glasses are not used in 3D video content, images may appear distorted. Two types of glasses, active and passive, are available depending on the Television receiver display. Active glasses are the most widely used type.

Active glasses used to watch 3D TV use shutter technology for each eye and communicate with the Television receiver via radio frequency waves or infrared light to synchronise with the display. These types of glasses need a power source to operate, often rechargeable batteries are used. Active glasses are expensive and are not interchangeable with different brands of television receivers.

Passive shutter displays require a different type of passive glasses which are cheaper and lightweight, but does not display a full HD resolution picture.

3D TV that do not require eye glasses to watch but gives a 3D display are also being developed and may be available in near future.

3D video content need to be made using special 3D cameras. Some 2D video content including movies have also been converted to 3D format using image enhancing technology but do not appear to give the true 3D display properties as much as original 3D content. Most of the 3D content are available in Blu Ray disc format. A 3D Blu Ray player with a 3D TV screen is needed to watch these content. There are several experimental 3D terrestrial Television broadcasts being telecast in many countries, in addition some cable and satellite channels also make 3D content available.

28

IPTV

IPTV is the abbreviation for Internet Protocol Television. IPTV is a digital television service delivered over a high speed connection which usually is an ADSL broadband connection. It is delivered by the IPTV service providing company over a managed network. In most instances the service providing company owns the network infrastructure as well. Video encoding methods like MPEG-2 or MPEG-4 and broadband delivery methods like ADSL2+ are used to deliver the IPTV channels.

Initially the speed of internet over ADSL was not sufficient to deliver good quality TV pictures but with the deployment of VDSL and ADSL2+ services it has been possible to deliver good quality IPTV services. It is also possible to browse internet while watching IPTV at the same time using the same connection. In some countries IPTV is delivered using fibre cables or a combination of fibre and ADSL.

In delivering IPTV, channels are encoded in IP (Internet Protocol) format and delivered to the TV using an IPTV set top box. The television channel signals are converted to small digital data packets similar to Internet traffic before delivering over the IPTV network.

There is a fundamental difference between cable TV and IPTV. In cable TV pictures are taken off a single cable which runs through a neighbourhood while in IPTV the connection to the telephone exchange from home is the dedicated phone line and is not shared with others.

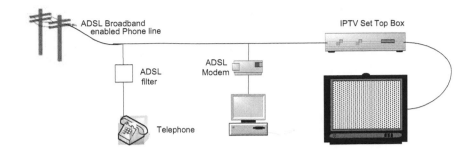

IPTV Set Top Box

An IPTV set top box is needed to watch IPTV channels. The IPTV set top box at the customer location is connected to the phone line and the video output is connected to a television receiver. IPTV set top box that receive IPTV data packets reassembles and decodes them to convert them back to a video signal. Some IPTV set top boxes also incorporate the function of an ADSL modem to connect to internet services. Wireless ADSL router functions are also available in some IPTV set top boxes where a computer can connect to the router through a wireless connection and access internet. IPTV services can be of both SD (Standard Definition) and HD (High Definitions) formats and a suitable IPTV set top box that can decode these formats is required to receive them.

In an IPTV network the channel selection is done through the IPTV set top box using the remote control unit. Once a channel is selected, only data related to that channel are sent to the receiver.

IPTV can also offer on demand TV where the viewer can choose to watch a program whenever they like. Some of the other features offered in IPTV are Electronic Program Guide, Interactive TV applications and targeted advertising.

IPTV set top boxes may also have the functionality of recording programs to an internal Hard Disk. Programs can also be recorded remotely on a server to watch later if the facility is provided by the service provider. Dual tuner IPTV set top boxes allow one program to be watched and another to be recorded at the same time.

127

Video on Demand

Video on demand gives the subscriber the freedom to watch a video at their convenience. On demand videos are usually available to be selected from a play list and can be watched whenever needed. On demand videos can be paused, skipped or backed up and watched.

Payment

To subscribe to IPTV an IPTV set top box need to be purchased and a subscription fee paid depending on the channels needed. Some channels may be available free of charge. On demand videos are usually charged on per movie basis.

Internet TV

Although the terms IPTV and Internet TV are used to mean the same thing, Internet TV is different from IPTV. Additional equipment is not needed to watch internet TV. Internet TV is watching a video stream over the public internet and is usually available free. Unlike IPTV, video streaming is not in a managed network and can affect the quality. In IPTV a certain QoS (Quality of Service) is maintained by assigning priority to packets carrying video. Watching a YouTube clip on a PC or logging on to a web site of a television station to watch a video stream of the television channel are examples of internet TV.

29

Cable TV

Cable Television is a television service that is distributed by a cable network. Radio frequency signals carrying television picture information are distributed through the cable network.

In a cable TV network' coaxial cable, fibre optic cable or a combination of the two is used as the medium for sending television signals. In some instances a fibre optic cable is used from cable companies to different neighbourhoods and signals moved in to a coaxial cable for distribution to individual homes. A cable TV set top box is needed to convert these signals for viewing on a television.

There are two different versions of cable TV, Analogue and Digital. In analogue cable TV analogue television channel signals are modulated and sent through the cable system.

Digital cable TV gives superior quality pictures and can provide SD (Standard Definition) and HD (High Definition) services. DVB-C (Digital Video Broadcasting-Cable) is the term used to denote cable based digital television services. MPEG-2 and MPEG-4 are used as video decoding schemes for digital television channels. The modulation system used in DVB-C is single carrier QAM (Quadrature Amplitude Modulation).

The cable TV network can also be used to provide internet and telephone services and these are offered as additional services through the cable TV network in some countries.

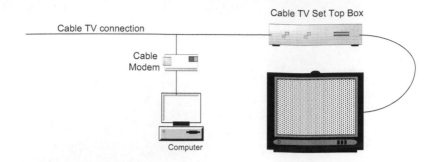

Cable TV Set Top Box

A cable ready TV is a television set capable of receiving cable television without a set top box. Only unencrypted channels can be watched without paying a subscription.

In most instances a payment has to be made for cable TV subscription and a set top box supplied by the cable TV company will decrypt the channels for which payment has been made. Most digital cable TV set top boxes are addressable type which means they have a unique address. A cable television company can send control information through the network to manage subscription services for individual set top boxes. The control signals will be sent throughout the whole cable network, but only the set top box with the intended address will pick up the control signals.

In most instances it is illegal to connect to a cable TV network without payment or obtaining permission. In some countries it is possible for anyone to connect to a cable network and watch free channels, a cable ready TV or a cable set top box can be used for this purpose.

Cable Modem

Some Cable TV distribution companies provide Internet access through the Cable TV network. Cable modems that allow high-speed access to the Internet via a cable television network can be used to access Internet. Cable TV modems are similar in some respects to an ADSL modem, but use the cable connection instead of the phone line.

DVB-C2 is the latest digital cable TV standard that is being developed.

30

Satellite Television

Satellite television reception involves the reception and display of television picture signals broadcast by a satellite. The geo stationary satellites that broadcast satellite television are positioned approximately 36000 km (22300 miles) above earth in a Geosynchronous Orbit directly above earth's equator called the Clarke belt named after Arthur Clarke. There are many satellites on this belt, some as close as 2 degrees apart. These satellites do not change their positions relative to earth since they follow the earth's rotation.

A satellite uplink station on earth transmits the satellite television signals to a satellite. The satellite act as repeaters of satellite television signals by retransmitting them to cover a wide area on earth where these satellite television signals can be received using a satellite dish antenna.

The advantage of satellite television is that unlike terrestrial television, satellite television signals can cover a vast geographical area. A large number of channels from different countries could also be viewed in satellite television

A dish type satellite television signal receiving antenna and a satellite TV receiver to decode the signals along with a TV set or monitor are required to watch satellite TV.

There are two main categories of satellite television signals, Analogue and Digital.

Analogue Satellite TV

Most of the earlier satellite television channels were transmitted in analogue format in the C Band and required large satellite receiving dish antennas. Analogue channels can be of PAL, SECAM or NTSC format. Due to new technological developments and the advantages offered in digital satellite TV, analogue satellite TV channels are gradually being phased off.

Digital Satellite TV

Most satellite TV channels are now transmitted in Digital format. It allows more channels to be offered for a given bandwidth. Digital satellite TV also offers better picture and sound quality than analogue satellite TV. A digital satellite TV receiver is required to receive digital satellite TV signals. DVB-S (Digital Video Broadcasting – Satellite) is the standard term used to identify this type of satellite TV. An enhanced standard which is an improved version of DVB-S will be used in future and is known as DVB-S2. Digital satellite TV signals use MPEG (Motion Picture Experts Group) format for compression.

MPEG-2/ MPEG-4

Most digital satellite transmissions use MPEG-2 compression format for digital signal encoding. Some satellite TV broadcasters are now starting to broadcast channels that use the newer MPEG-4 format. MPEG-4 format takes less bandwidth and can offer more channels. Most High Definition channels are transmitted using MPEG-4 decoding. To view satellite channels encoded with MPEG-4 the satellite receiver should have MPEG-4 decoding feature.

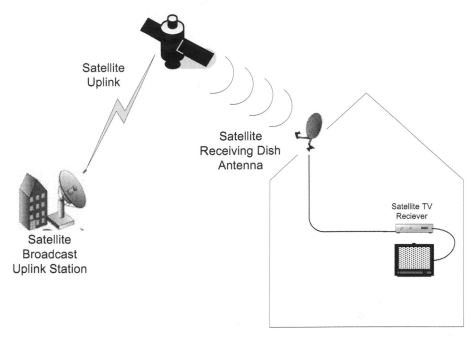

SD/HD

Digital satellite TV signals can be of SD or HD type.

HD (High-definition) is a digital video format that can display images that are much sharper and clearer than SD (standard television).

A High-Definition TV (HDTV) picture comprises of almost double the number of lines than a Standard Definition (SDTV) picture. The picture will have more details since it contains more lines.

HD broadcasts provide a rectangular widescreen shape for viewing. It also has Dolby Digital 5.1 Surround Sound facility.

Satellite television receivers come with SD or HD tuners. A SD type receiver decodes the SD type digital broadcast signal and sends it to the TV. An HD satellite TV receiver is designed to decode high-definition satellite signals, and it sends them to HDTV or HD-ready TV in widescreen digital format. All satellite channels are not broadcast in HD. Usually a HD satellite receiver can decode SD channels as well. To view HD satellite TV channels decoded from a HD satellite TV receiver in true HD format the television receiver should be an HD type TV.

Satellite Dish Antenna

A satellite television receiving dish antenna can be as small as 45cm across, or it can be 3m or larger across and parabolic in shape. The purpose of the dish antenna is to act as a collector of satellite TV signals and to reflect them to its focal point. At the focal point of the satellite dish a feed horn (similar to an antenna) collects the reflected signal. A larger dish antenna is required to gather weaker satellite television signals when the receiving antenna is located further away from the centre of the satellite footprint

A ku band satellite dish antenna   **C band 2m diameter**

satellite dish antenna

LNBF (Low Noise Block Converter Feed horn)

LNBF is the device on the front of a satellite dish. It receives the very low level microwave signal from the satellite which are collected and reflected by the satellite receiving dish. LNBF amplifies it, converts the signals to a lower frequency band and sends them to the satellite receiver via a cable. These converted signals usually use 950-1450MHz frequency range and are sent in the cable in a band called the L band.

The signals are converted to a lower frequency band to reduce the losses when they are sent through the cable to the satellite receiver. In higher satellite frequencies where satellite TV signals are broadcast the signal losses in cable are quite high. Most of the time the DC power required to operate the LNB is sent from the satellite receiver unit

134

through the same cable that LNB sends the down converted satellite TV signal.

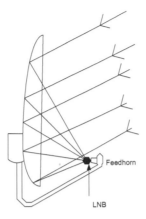

Signals from satellite are reflected to the LNB by the dish antenna

Satellite Dish alignment

The satellite dish has to be aimed at the satellite where the desired TV signals are transmitted. If the dish is a few centimetres away from the intended direction the signals from the desired satellite will not be properly received. Satellite dishes are highly directional and the larger the dish the more precise it has to be aimed. The direction of aiming the satellite antenna for a particular satellite is different from different locations on earth and has to be calculated depending upon where the antenna is located. Since the Clarke belt where satellites are positioned is located above the earth's equator, satellite receiving dish antennas in Northern hemisphere have to be directed Southwards, antennas in Southern hemisphere have to be directed Northwards and antennas near the equator have to be directed upwards. Sometimes the dish may not seem to be directed upwards but depending on the focal point where the feed point is located it will be positioned to collect signals arriving from upward direction.

Azimuth and Elevation

The dish antenna has to be aimed according to the calculated azimuth and elevation from a particular position on earth. Azimuth is direction relative to magnetic north which is also known as left-right direction and a compass is used to align the dish to the correct Azimuth.

Elevation is the angle between the Earth (horizontal plane) and the direction of satellite dish or commonly known as up-down angle. Usually the satellite antenna mounting bracket has a scale marked indicating the angles.

There are computer software to calculate the azimuth and elevation for various satellites from a given location. The geographic coordinates (Latitude and Longitude) of the position of receiving antenna is entered and the result gives the azimuth and elevation to point the dish antenna for a given satellite. Satellite pointing charts for various locations can also be used to aim dish antennas. With some satellite receivers used in some countries it is possible to enter the postal zip code of the receiver location and obtain the antenna pointing directions.

Tilt

If the satellite dish antenna is equipped with more than one LNB to receive signals from multiple satellites, the tilt also has to be set up.

The satellite dish has to be mounted in a position where there is unobstructed view of the sky.

The received satellite signals are quite weak since they have travelled over 36,000kms from the satellite to the earth. The dish antenna is used to collect the weak signals over a large area which is the surface of the dish. As the dish size becomes larger, it can gather more signals.

Usually the satellite dish is fixed to receive signals from one satellite only. If signals from more than one satellite are to be received there are various arrangements for that purpose.

Multi satellite reception

Two or more dish antennas aimed at different directions where satellites are positioned can be used for multi satellite reception. The signal from the desired dish antenna can be switched on at a time. To avoid using multiple dishes there are other techniques to receive signals from more than one satellite.

A single dish antenna can be used with multiple LNBs each angled differently at different focal points of the dish. These multiple LNBs can receive signals coming from different satellites at different angles which are reflected to the different focal points. A switching mechanism is used to select the desired satellite signal.

A more complicated method is to use motorised steerable dish antenna system. A dish antenna mounted on a motor is steered by a known amount according to instructions given from an antenna controller (positioner). Positioner can be a separate unit or part of a receiver and it sends control information to the motors to steer the dish.

There are two types of positioning used. In one type there are separate cables running from positioner to the motor that carry motor control information. Other type is where all motor control information is carried in the same coaxial cable that also carries the TV signal and is connected to the LNB and receiver.

A satellite TV receiving dish with multiple LNBs to receive signals from several satellites

Satellite Television Bands

C Band

C-Band range is from 3.7–8 GHz while C-Band satellite TV broadcasting uses 3.7-4.2GHz frequency range. C Band satellite dishes can vary from 1.8-3.7m in diameter, typical ones are about 2m diameter. C Band satellite TV signals perform better under adverse weather conditions compared to Ka and Ku bands because the frequency used is lower. Most of the early day analogue satellite TV channels used the C band.

Ku Band

Kurtz band also known as K band. Ku band range is from 12-18 GHz while 11.7-12.7 GHz frequency range is used for Ku band satellite TV broadcasting. Ku band signal are susceptible to degradation due to rainfall because of their higher frequency. Most digital satellite channels are now transmitted in the Ku band. Smaller dish antennas, typically of 0.3m–1m diameter can be used for Ku Band. Ku Band signals are transmitted at a much higher power than C Band from the satellites. It is possible to receive Ku band signals with smaller dish antennas due to the increased transmit power at satellites and the higher frequency band used.

Satellite footprint

Satellite footprint is the area on ground that a particular satellite signal can be received with reasonable strength. Different satellites have different satellite footprints to cover various geographic regions on earth. To receive signals from a given satellite the receiving dish antenna has to be within the satellite footprint area of that satellite. Satellite TV signals received in the centre region of the footprint area are usually much stronger requiring smaller dishes than signals received in the border region of the footprint area where signals may be weaker and larger dishes may be required.

Free to Air (FTA) satellite TV channels

FTA satellite channels can be received with a FTA satellite TV receiver and these channels are available free of charge. These FTA satellite channels are transmitted in unencrypted format. There are hundreds of FTA channels available and it does not cost anything other than the initial equipment cost to receive these channels.

Encrypted/scrambled channels

Encrypted satellite TV channels can be received and decoded by a suitable satellite TV receiver that can decode encrypted signals. Encryption is used to limit the viewing of the channels only to paid subscribers. Most satellite receivers have a slot to insert a smartcard to decode channels which are encrypted for which payment has been made for viewing. Each satellite receiver has a unique address stored in it. The satellite broadcaster can use this address to send a signal through the satellite to decode the channels for a particular period for which payment has been made.

Many techniques are used for encrypting satellite TV channels. Irdeto is a popularly used such technology for encrypting digital channels. Nagravision, viaccess, Wegener, Powervu are some of the other digital encryption methods.

Satellite Radio Channels

There are a few Audio only (without pictures) channels that are being broadcast on the satellite television frequency bands, most of them are of Free To Air type. These types of satellite audio channels are known are DTR (Digital Television Radio). These can be received with Satellite TV receivers. (There is a separate satellite radio service which use portable satellite radio receivers which is explained in a separate chapter)

Satellite TV Receiver

A satellite TV receiver converts the received satellite TV signal to a format that can be watched on a Television set. It also extracts one channel at a time from many satellite TV channels received.

There are various types of satellite receivers. These receivers can receive analogue signals, digital Free To Air channels, or encrypted digital channels. There are also receivers which has a combination of these functions. The receiver has to be chosen to receive the required band C, Ku or Ka bands. There are receivers that can receive multiple bands.

A satellite TV receiver

If HD channels are to be viewed the satellite receiver should have HD decoding features. The compression type (MPEG-2/MPEG-4) that needs to be decoded also has to be considered when selecting a receiver. Newer receivers can decode MPEG-4 channels. MPEG-4 is an improved picture compression standard and takes less bandwidth than MPEG-2 to give a high quality picture.

For digital channels the correct settings for the channel such as frequency, polarisation (horizontal or vertical), symbol rate (SR) and the FEC (Forward Error Correction) have to be set in the receiver. Satellite receivers sold directly by the satellite broadcasters may have been pre-set up with the settings for the required channels. Sometimes these receivers may be locked to prevent the reception of channels from other satellite broadcaster, or even Free To Air channels.

Some satellite receivers have a telephone jack for a connection to the telephone network to communicate with the satellite broadcast control station.

31

Facsimile (Fax)

Facsimile which is commonly known as Fax or Telefax is the transfer of an image or copies of documents over a telephone line.

A fax machine combines the function of an image scanner, a printer and a modem. The image of a document fed in to a fax machine is scanned and converted in to digital signals. These digital image signals are converted to analogue telephone line signals by the modem in the fax machine and sent via telephone lines to the destination fax machine. The received signals are converted back in to an image and printed by the destination fax machine.

A fax machine

Plain paper and Thermal Paper fax

Older fax machines used Thermal paper for printing. Thermal paper could usually be purchased as a roll and was coated with a chemical that turns black due to the applied heat. Thermal paper fax machines do not require separate ink cartridges for printing. Thermal paper fax machines are less expensive than plain paper fax machines as the printing mechanism is simpler. The disadvantage with thermal paper is that the printing fades over time, specially when exposed to light. Therefore to preserve the received images for a longer period it is required to photocopy them on to plain paper.

Newer fax machines use plain paper for printing. The printing mechanism is similar to that of a computer printer and most often are inkjet, bubble jet or laser type.

A plain paper fax uses the same paper that is used in a photo copier or laser printer to print out faxes.

Colour Fax Machines

With the integration of computer printing technology to fax machines, it is now possible to fax colour documents between Fax machines that support colour faxing. The relevant colour fax standard ITU T.30e was adopted recently but is still not used widely. It is usually possible to send colour faxes between same brand fax machines that support colour faxing even though they may not be compatible to ITU colour fax standards. To send colour faxes between different brand fax machines they should be compatible with ITU colour fax standards

Faxing using dialup modems

A computer and a dialup modem that support fax can also be used for sending and receiving fax using the appropriate fax software. For example, a word document on a computer can be sent by fax using the computer dialup modem (internal or external) without having to use a fax machine. If an image on a paper has to be faxed, it can be scanned using a computer scanner and faxed using the computer modem.

Group 3 Fax Standard (G3)

Group 3 fax machines are the current standard fax device for sending fax over analogue phone lines. Group 3 fax machines conform to ITU-T Recommendations T.30 and T.4.

A Group 3 fax machine should have another Group 3 fax machine to communicate at the other end. It takes between 6-30 seconds to send one page in addition to the initial 15 second handshake period for a Group 3 fax machine. Handshake is the initial communication that occurs between devices in order to determine the method and speed of data transfer to be used. Data transmission speed and the time to send a page depend on the fax machine used and the telephone line conditions.

Newer machines can send faxes at up to 33.9kbps data transfer speed.

When two fax machines that support different data transfer speeds communicate, their data transfer speed will be the maximum speed of the slower machine.

Group 1 and 2 fax machines which were in use before Group 3 fax machines took 3-6 minutes to send a page and are rarely used now.

Group 4 Fax (G4)

Group 4 fax machines are digital fax machines that are used to send faxes over ISDN telephone lines. Time taken to send a page using a Group 4 fax machines is about 4 seconds. Data transmission speed in a Group 4 fax machine is up to 64kbps

Fax Machine Features

Distinctive Ring detection Feature

Distinctive Ring is a service provided by telephone companies in some countries. It enables two separate phone numbers for the same phone line. Each number makes the phone ring differently. One number can be used for ordinary voice telephone calls while the other can be used for fax. When the voice phone number is dialled the phone rings normally and the recipient can pick up the call. If the fax number is dialled the phone rings with a different ringing tone indicating it is a fax. Some of the fax machines support distinctive ring feature. These machines can distinguish the two different ringing tones and will answer the fax ringing tone only. Only one of the numbers can be used at one time since a single phone line is used. This feature is different from Auto Tel/Fax feature described below

Auto Fax/Tel feature

The auto fax/tel feature allows the use of single telephone line and single phone number to receive both fax and voice telephone calls. The fax machine can be programmed to pick up a call after a number of rings and if a fax tone is detected from the remote end the machines will do the handshaking and begin transferring the fax. If a fax tone is not detected the fax machine will continue to ring to indicate that the call is a voice call and not a fax and also send a fax machine generated ringing tone to the remote party until the handset is picked.

Resolution

This feature allows adjusting the quality of the scanning of the document. Plain text needs standard resolution, while graphics may require fine or superfine resolution. Higher resolution documents take longer to send and produce finer prints.

Remote Reception

Remote reception allows the use of an external phone connected in parallel as an extension to the same phone line to enable the fax machine to receive a fax. This is useful when the fax machine is not nearby to the phone. When a call is received and the phone extension is picked up, if the received call is a fax and gives a fax tone, dealing a short code (typically 2 digit remote reception ID) on the phone will manually enable the fax machine to receive the fax.

Delayed/Scheduled transmission

Some fax machines enable faxes sent at a scheduled time or a later time. This feature is useful to take advantage of off peak cheap phone rates. The documents are scanned and kept in memory of fax machine and will be sent at the programmed time without the need of user intervention.

Out of paper Reception

This feature allows a fax machine to receive faxes even when there is no paper in the machine. The received fax is saved in fax memory and printed when paper is loaded. However if power supply is interrupted to the fax machine before the saved fax is printed the saved fax may be lost.

Paper cutter

This is a useful feature in fax machines that use thermal paper rolls. Paper cutter will automatically cut each received page from the paper roll at the end of reception.

Copying function

Most fax machines can be used as a copier also. A document page can be scanned and printed using the fax machine.

Confirmation page

This is a useful feature to keep track of successful and unsuccessful transmission of faxes. A confirmation page can be printed for each individual fax which shows the date and time of transmission, number of pages and time taken for transmission and whether the transmission was successful. If the transmission was unsuccessful the reason (recipient line busy, no fax tone at remote end etc.) is also usually given. This feature can be turned off if required to save paper.

Fax log/Activity Report

This feature will print a list of all faxes sent and received and is useful to keep track of all faxes sent and received through the fax machine.

Broadcasting

Broadcasting allows a fax scanned once to be sent to multiple fax numbers.

Group Dialling

This is similar to fax broadcasting. If faxes need to be sent to a particular group frequently, the fax numbers of recipient group can be pre-programmed to the fax machine.

Dual Access

This allows a fax machine to print incoming faxes from memory while transmitting another fax, or scan a fax into memory while receiving a fax.

Quick Scan

Quick scan will quickly scan a fax into fax machine memory before a fax is transmitted, and then the fax is transmitted from the fax machine memory. It takes less time to retrieve a page fed into a fax machine.

Polling

Polling enables a fax machine to call another fax machine and retrieve a document from the remote machine.

Polling – Sending

Allows sending a document in response to remote machine's request.

Polling – Receiving

Allows retrieving documents stored in remote machine for polling.

Relay Initiation

Allows sending a document to a remote machine (relay station) which in turn sends the document to other destinations depending on the numbers programmed in the relay station.

User information setting

The user's or company name and phone line number can be programmed in the fax machine which is then transmitted along with every fax. The remote party can see these details on top of received fax page.

32

Trunked Radio

Trunked Radio is a Radio communication system used for commercial and utility service purposes. It is also known as Trunked PMR (Private Mobile Radio) or LMR (Land Mobile Radio). It is used by Public Safety firms like fire brigades, police, paramedics, transportation firms like taxi companies, utilities services such as electricity and water service personnel, Commercial & Industrial firms, Oil & Gas companies, Sea Ports, airport services and Military.

Conventional radio communication services use a dedicated frequency channel for communications. These channels are used only when needed and they remain unused most of the other times.

In trunked radio communication all users share a pool of frequency channels and use channels only when they are needed. Trunked Radio technology was developed to make efficient use of all frequency channels evenly.

A handheld unit and mobile unit used in Trunked radio

A trunked radio communication system consists of one or many base transmitter/receiver stations and repeaters which are connected to a control station that assigns frequency channels to talk groups. User terminals are handheld radio units or mobile units fixed in vehicles. The control station keeps track of all mobile unit ids, their status and operational conditions. The base stations and repeaters which are controlled by the control station receive the signals sent by user units, amplify them and retransmit them over a wide area.

In a trunked radio communication system there are many groups of users. A user can use his communication set to communicate with members of the same group. Users are assigned a unique group id and group member radios are programmed to pick up transmissions intended only for that particular group.

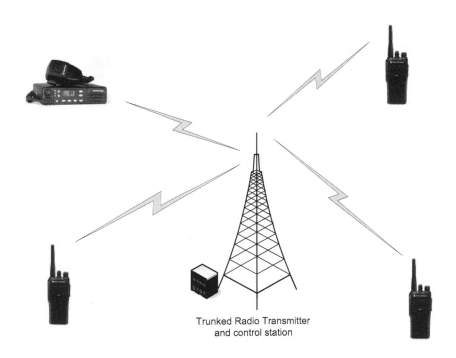

Trunked Radio Transmitter
and control station

All radios listen on their control channel (data channel) in the idle mode. Whenever a member of the group needs to talk to others in the group the PTT (Push to talk button) on the radio has to be pressed. Then the control station assigns an unused frequency channel for users belonging to the same group to communicate with each other. A control signal is sent in the control channel to all radios in the group to switch to the assigned frequency channel for communication. This process takes fraction of a second and is digitally controlled automatically by the control station. The user units switch to the assigned frequency for the duration on the conversation.

At the end of the conversation the frequency channel is released and the radios get back to the idle mode and monitor the control channel. The released frequency channel is now available to be assigned to any other group (or even to the same group) when requested.

There are many advantages of trunked radio systems over conventional systems. Since the frequency channels are shared among many groups of users and are in use most of the time there is efficient usage of the limited radio frequency spectrum. A commercial trunked communication system operator builds the infrastructure for the system to cover a wide area and the customers can rent equipment for their use from the trunk system operator. This avoids the need for individual users having to build their own systems and to have dedicated frequency channels. Different companies can rent communication sets for their individual groups and a group can range from two users to a large number of users. Different utility services also operate their own trunk radio networks limited to areas where they operate; examples include airports and sea ports.

Analogue and Digital Trunking systems

There are both analogue and digital trunked radio systems in operation. In analogue trunked system the communication voice channel signals are sent in analogue form but in digital systems voice channel signals are encoded in digital format.

Tetra

Tetra is the abbreviation for Terrestrial Trunked Radio. It is a digital trunked radio system standardised in Europe and used in many parts of the world.

It offers many advanced features such as

- Individual call – connects one user of the group with another single user in the group

- Dynamic group – users in a group can be changed dynamically, group do not need to be fixed

- Broadcast call – A call sent from control station to all users

- Direct mode – two users within range can talk to each other directly without going through the system

- Emergency call – Priority channel assignments in emergencies

Tetra also facilitates both voice and data communications at various data rates.

P25

P25 is a set of standards for public safety communication services. It was developed primarily for North American Public Safety Radio services, but has been deployed worldwide in other private system applications. P25 Digital Radios can communicate with earlier analogue two way radios and with other P25 radios either in digital or analogue mode. P25 can be used in both trunked and conventional two way communication modes.

33

Cordless Telephones

A cordless telephone or portable telephone is wireless handset which communicates with its base unit connected to a fixed telephone line via radio waves. A cordless phone enables one to talk on the phone using the fixed phone line connection while moving freely about in the house or in the garden. It can only be operated close to (typically less than 100 meters) its base unit.

A cordless phone has two major parts, a base unit and a cordless Handset.

Base Unit

The Base unit is connected to the phone line through a standard phone wire connection. The base unit receives the incoming call through the phone line, converts it to a radio signal and then transmits that signal to the handset. It also receives radio signals from handset and converts them to be sent through the phone line. The base unit also functions as the charging unit for the cordless unit and requires to be connected to mains power all the time.

Cordless Handset

The handset receives the radio signal from the base unit, where it is converted into speech. When we talk, the handset transmits our voice through a second radio signal back to the base unit. The base unit receives the radio signal, converts it to an electrical signal and sends that signal through the phone line to the other party. The handset has a rechargeable battery and when the battery is low it should be placed on the base unit for the battery to recharge.

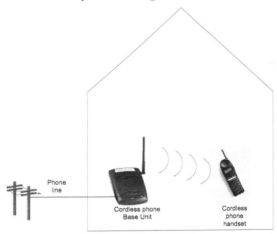

Phone line

Cordless phone Base Unit

Cordless phone handset

Power Supply

A cordless telephone requires household mains electricity to operate the base unit. The system is not operational during power failures. The cordless handset is powered by a rechargeable battery. This is a nickel-cadmium (NiCd), nickel-metal hydride (NiMH) or lithium (Li-Ion) type battery. When the battery runs low, an indicator light on the handset usually lights up or flashes. In some phones, a warning sound may also indicate a low battery. The handset battery can then be recharged by placing the cordless handset on the base unit for several hours.

The battery may become weak after several years of reuse and may not hold the charge. The rechargeable battery then needs to be replaced with a similar type rechargeable battery.

Analogue/ Digital cordless phones

Analogue transmission technology is common in cordless telephones, especially in inexpensive models and older cordless phones. Analogue signals give reasonable voice quality though at times tend to be noisy, or prone to interference and static noise. As an analogue telephone handset is moved further away from its base unit, its signal fades and we hear radio static noise. Digital cordless phones offer better signal quality than analogue cordless phones. Digital telephones convert the voice into a series of binary (1's and 0's) numbers and transmit those numbers. The receiver converts the binary digits to voice signals. The circuitry in a digital receiver has only to detect whether the received signals contain 0's or 1's. With advanced digital error correction mechanisms digital cordless phones can provide better signal quality than analogue cordless phones. The result is that the digital phone's handset will operate at a fairly good distance from its base without any interference. Digital cordless phones provide a stronger, clearer signal and improved security. Digital transmission is less prone to interference from radio waves.

Cordless phone Radio Frequencies

The base unit and handset operate on a closely related but separate frequency pair (duplex frequency) that allows one to talk and listen at the same time.

Channels

Cordless Phones operate on different frequency bands depending on each model.

- 43-50 MHz band for analogue cordless phones

- 900 MHz band analogue cordless phones which are only allowed in certain countries.

154

- 1.9 GHz Band

- 2.4GHz and 5.8GHz bands for digital cordless phones.

Each frequency band can be subdivided into different channels. The base unit searches for a pair of frequencies (channels) within the allocated frequency band that is not already in use, in order to talk to the handset. If there are more channels available it can more easily find a frequency pair that is clear from interference, providing better sound quality.

The number of cordless phone channels available can vary depending on the frequency band of the cordless phone.

A conventional cordless phone is designed to use only one channel (or radio frequency) at a time. Fixed frequency cordless phones will not switch to a new channel unless they encounter interference or if the user forces a channel change manually by pressing the channel button on the handset if available.

Digital cordless phones use two modulation technologies for signal transmission. They are DSS (Digital Spread Spectrum) and FHSS (Frequency Hopped Spread Spectrum).

Cordless Phone Security Codes

Most cordless phones incorporate a security code to prevent outsiders from accessing the phone line, and placing unauthorized phone calls. Security code ensures that only cordless phone handsets belonging to the base unit can make calls through the base unit and connected phone line. Handsets belonging to other base units are prevented from accessing the phone line. This security level is provided by a digital security code that is randomly generated from about 65,000 possible codes each time the cordless handset is placed on the base unit. Sometimes when the cordless phone unit does not work, placing the handset on base unit will reset the security code and this will clear the problem most of the time.

Features in cordless phones

Convenient features make the cordless phone easier to use. Some of these features are identical to those in standard corded phones like Calling Line Number Display, Speakerphone, Mute/Hold, Volume control, Ringer on/off and volume set.

Some Convenience features found in cordless phones:

Two-Line Service

This feature allows two separate phone lines to be connected into the same cordless phone. Some of these phones have two different handsets, while others use one handset which can select the phone line to use. This feature has the advantage of using one phone for two lines rather than having two separate phones.

Two/Three handsets

This enables a single base unit connected to one phone line to have two or three cordless handsets.

Pager/Handset Locator

This feature is most helpful in locating the handset when it has been misplaced. If this feature is available, a button can be pressed on the base unit which sends a radio signal to the cordless handset, When the handset receives the radio signal it sets off a beeper enabling to locate the handset by the sound of the beeper.

LCD Screen

LCD screen on either the handset or base unit can display useful information including name/number dialled, battery strength, saved phone directory name and number. Incoming phone number can also be displayed on handset or base unit if the service has been enabled by the phone company.

Out of Range indicator

When the cordless handset is moved far away from the base unit and out of range of service, this feature will give an indication by a flashing light and/or warning tone.

Low Battery Indicator

When the cordless handset battery is low a flashing light and/or a warning tone will indicate that the battery needs recharging by placing the handset on the base unit.

Ringer On/Off /Volume

This feature can disable the ringer and also adjust the ringer volume to one of the several levels.

34

GPS

GPS is a worldwide satellite radio navigation system. It enables users with GPS receivers to know exact position details (longitude, latitude and altitude, position in a map) they are in. GPS technology is changing the way how people find their way around the world. GPS navigation is increasingly being used in various situations such as to find the way after being lost in an unknown territory, in vehicles to find the optimal route, in aviation and ships to know the exact position etc.

GPS is the abbreviation for Global Positioning system and is officially known as NAVSTAR GPS (**Nav**igation **S**ignal **T**iming and **R**anging **G**lobal **P**ositioning **S**ystem).

GPS consists of 24 GPS satellites that orbit the earth and transmit radio signals. GPS receivers that receive these signals can indicate their locations, speed of travel, direction, time etc. GPS system was developed by the United States Department of Defence and is controlled and maintained by the US Air Force.

GPS consist of 3 main segments. They are the Space segment, consisting the Satellites, the Control Segment consisting the Ground control stations and the user Segment consisting GPS receivers.

Space segment

24 GPS satellites (plus 3 spare satellites to be used in case of failure of a satellite) orbit the earth at an altitude of 20,200km. The

satellites are positioned in a way that at any given moment at least 4 satellites are within line of sight of any place on earth.

Each satellite travels at a speed of 11,300km an hour and circle the earth once every 12 hours. The satellites are powered by solar energy with backup batteries and are equipped with small rocket boosters to keep them in correct path. The satellites transmit low power (less than 50W) radio signals. Each satellite weighs around 1000kg and is about 5m across in size with the solar panels extended.

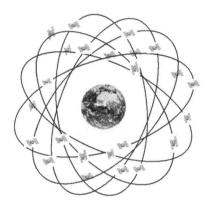

24 GPS satellites orbit earth in 6 different orbital planes.

Control segment

The control segment controls the GPS system by tracking the satellites and sending control signals to them for correct operation.

There are 5 control stations. The main control station of GPS system is located on Schriever Air Force Base Colorado USA. Four other unmanned tracking stations are located in Hawaii, Kwajalein (Marshall Islands), Ascension Island (South Atlantic Ocean), Diego Garcia (Indian Ocean) and Colorado Springs (USA). These stations monitor the GPS satellites and send the information to the main control station.

User segment

The user segment consists of GPS receivers and the users of them. This include the military, motorists, hikers, ships, planes and whoever using a GPS receiver to know their exact location.

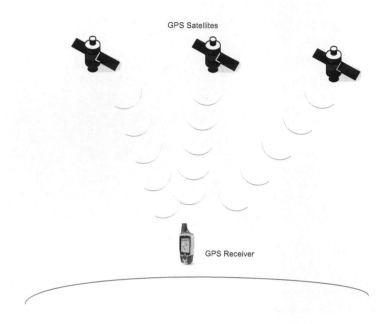

GPS Satellites

GPS Receiver

GPS satellite Signals

Each GPS satellite transmits radio signals on several frequencies. The signals are in UHF frequency band where they propagate line of sight which means that they can pass through clouds, glass, plastic etc. but are blocked by solid objects like buildings and mountains.

Most widely used GPS signals for civilian use are known as L1 and L2.

- L1 – 1575.42 MHz
- L2 – 1227.60 MHz

GPS position calculation

A GPS receiver's function is to receive signals from three or more GPS satellites, calculate the distance to each of them, and use this information to calculate its own location. If 3 satellites are visible a 2D (2 Dimensional – latitude and longitude) position can be calculated while if 4 satellites are visible a 3D (3 Dimensional – latitude, longitude and altitude) position can be calculated. The more satellites the GPS receiver can see the better the accuracy of position calculation. Usually 5-8 satellites are visible most of the time. (The term visible used here means the satellites are in line of sight, not visible to the naked eye).

A GPS receiver contains in it, information about the exact positions of GPS satellites at all times. It knows the signals it should receive from each satellite at any time. GPS receiver software can identify the transmitting satellite from the signals received and its location in space and calculates the time it took to the signal to travel from satellite to the receiver.

GPS receiver using its internal clock compares the timing of actual signals received from satellites and the signals it should have received from visible satellites. From the delay between these two it can calculate the time the signal to reach the GPS receiver from satellite. From these data it can calculate the distance to each visible satellite. Using individual signals from each visible satellite the GPS receiver can calculate the position of the receiver. This calculation is based on a mathematical technique called trilateration. Trilateration uses distance to known reference points from a point to calculate the location of that point.

GPS Receiver

A GPS receiver has a number of receiver circuits to receive satellite signals simultaneously. The number of receiver circuits is identified as the number of channels of that GPS receiver. Older GPS receivers were of single channel type, modern ones have 5 or more receiver circuits (channels) with each one capable receiving separate satellite signal.

Most of the latest Smart Mobile phones and tablets now have a GPS receiver built in.

GPS receivers usually do not work properly indoors or in places where the satellite signal is blocked due to some obstruction.

A handheld Garmin GPS   **Trimble GPS**

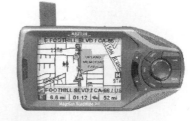

A GPS unit for vehicles   **Magellan GPS showing area map**

Tracking

A user with a GPS device can keep track of one's path and save it in the GPS device.

The current location coordinates can be used with a reference map to determine the exact location. The saved data can be used to follow the same path again later.

Timing information

GPS receivers synchronise with the highly accurate atomic clocks in GPS satellites and can provide a highly accurate time signal. These precise timing signals are used by power companies, computer networks, communications systems, and radio and television stations to synchronize their communication networks.

GPS in vehicles

When installed in a car, a GPS unit can provide useful information about the car's location and the best travel routes to a given destination by linking itself to a digital map. The driver can enter the target location address, and the GPS will calculate the optimal route and display it instantly. It can show the names of roads and best travel routes to a desired location. If the map is detailed enough, it can also provide the locations of the nearest gas stations, supermarkets, restaurants, hotels, and ATM machines. Most GPS units can issue instructions in voice commands (i.e., "Turn left,") to guide drivers as they travel.

GPS is quite a useful tool for motorists. It can track the distance travelled on a particular trip, vehicle mileage, and speed. A particular route taken can be saved in GPS memory for later use if desired. Some advanced GPS units offer the facility of sharing route and traffic information obtained from many users to calculate the optimum path depending on road and traffic conditions. Some GPS receivers have the ability to receive traffic updates (when the service is available) and see the route congestion and also to alter the route avoiding traffic congestions.

In many countries local digital maps are incorporated in to GPS devices. It can show the relevant portion in a map where the vehicle is located. Many models of modern vehicles now have GPS devices installed as an option.

Vehicle tracking using GPS

GPS receiver together with a mobile communication transmission system can be used to remotely keep track of one or more vehicles. A vehicle with such facilities can transmit its location through a mobile phone network in real time and the results can be displayed in a map. Alternatively recoded data can be used later on to track the vehicle movements with time, speed and location data. Many commercial systems are available to keep track of vehicle fleets such as taxis, police and emergency vehicles, vehicles in a distribution fleet, military vehicles etc.

Tracking applications for mobile phones

It is possible to install tracking apps in smart phones equipped with GPS and report the location of the phone user when the service is activated. This feature is quite useful to keep track of locations of children and elderly people who can carry a smart phone with the required application enabled.

Military use

GPS system was initially developed for US military use but later it was opened up for civilian use as well. US military use GPS in missile warheads to accurately target them to the required locations. Cruise missiles are equipped with GPS receivers to accurately hit targets from a large distance. These missiles constantly use GPS signals while in flight to guide its path to the target. Soldiers, battle tanks, convoy movement rely on positional data by GPS to determine the required locations.

The US Military use a different signal from GPS satellites which is more accurate than the signals received by civilian users. These signals are designed to be used only by US military.

Selective Availability (SA)

Selective Availability is a feature in the GPS that introduces an intentional error of 0-100m for civilian users. This was done by US military to prevent enemy forces from using GPS signals to target missiles accurately. Accurate information was only available in encrypted form to US military and its allies.

Selective availability was turned off in year 2000 but can be reintroduced any time if required. Discontinuation of SA resulted in civilian and commercial users being able to obtain very accurate position data using GPS receivers.